## Praise for *The Invention of Air*

"Johnson is an exemplar of the postcategorical age. . . . [His] new book, *The Invention of Air*, shows its genre-mixing in its subtitle; it uses Priestley as the fulcrum for a story that blends 'science, faith, revolution, and the birth of America.' What enlivens the book is that Johnson does not simply describe the system within which Priestley and his contemporaries hashed out the features of classical science; he sets it against other, later systems for comprehending physical reality, showing laymen how far we have come from the classical age of science. . . . The 'long zoom' approach gives Johnson's book power, makes it a tool for understanding where we stand today, and makes it satisfying. . . . Brilliant."  —*The New York Times Book Review*

"Exhilarating . . . Like Priestley, Johnson—who wrote the bestselling *Everything Bad Is Good for You*—is a polymath, and . . . to explain why some ideas upend the world, he draws upon many disciplines: chemistry, social history, geography, even ecosystem science."  —*Los Angeles Times*

"Steven Johnson is that rarest of commodities among twenty-first-century public intellectuals: a progressive—both in the old-fashioned sense and the synonym-for-liberal sense—and an optimist. . . . His is a questing, limber intelligence, eager to consider opposing arguments, explore new terrain, and notice underlying patterns he hasn't seen before. I don't know whether it was God or video games that made him so smart, but something did. . . . [He is] an infectiously exciting writer [and] *The Invention of Air* is delightful to read. But it aims high. It isn't a work of conventional history or biography, though it contains snippets of both, but more like a case study in the history of ideas that hints at a grander analytical theory. Johnson is a wide-ranging enthusiast with a catholic appetite for intriguing facts and a Marxian appetite for searching for structures that underlie social phenomena . . . It's a good time for all Americans to imbibe a little of Johnson's, and Priestley's, irrepressible hopefulness."  —*Salon*

"This is not a book about the discovery of oxygen but about the invention of air: how groups of scientists, natural philosophers, religious leaders, and politicians served as cultural petri dishes in which ideas were discussed, experimented with, discarded, or accepted. . . . [Johnson] gives long-overdue time and space to some of the more controversial aspects of [Priestley's] work. . . . Priestley may not have gotten full credit for his work on oxygen, but this new book gives plenty to the life of the man himself."  —*The Dallas Morning News*

*continued . . .*

"The author of *Everything Bad Is Good for You* provides an entertaining account of the eighteenth-century scientist and radical Joseph Priestley's monumental discovery that plants restore 'something fundamental'—what we now know as oxygen—to the air. Johnson also offers a clear-sighted and intelligent exploration of the conditions that are propitious to scientific innovation, such as the availability of coffee and the unfettered circulation of information through social networks." —*The New Yorker*

"We rarely hear of [Joseph Priestley] today, but it wasn't always thus: the correspondence between Thomas Jefferson and John Adams includes fifty-two mentions of Priestley, versus just three of George Washington. With *The Invention of Air*, Steven Johnson brilliantly explains why. . . . For all of Priestley's many achievements, laid out so delightfully in Johnson's account, it's his work with plants and the oxygen cycle that rightfully gained him immortality. . . . Engrossing." —*The Oregonian*

"Steven Johnson's mind works in wondrous ways and readers have been the beneficiaries of his eclectic interests. Johnson's new book, *The Invention of Air*, marks a return to cultural history. . . . His free-ranging mind and irreverent wit entertain and prompt thought." —*Seattle Post-Intelligencer*

"Steven Johnson's latest book, *The Invention of Air*, is a wide-ranging, learned, engrossing biography of the polymath pioneering scientist Joseph Priestley. . . . Johnson uses the life of Priestley to illuminate a theory of history that holds that great people are neither an inevitable product of their times, nor luminous, supernatural geniuses—rather, they are the product of an *ecosystem* of influences, technologies, climate, and energy (literally—the story of stored energy in coal, saltpeter, and plant-bound carbon are vital to the story). He pulls this off deftly, with a series of insightful, beautifully realized anecdotes from the life of Priestley and his contemporaries—his allies and his many enemies—that make the idea of history being shaped by webs and networks seem absolutely true." —*BoingBoing*

"[Johnson] refracts just about every beam of Enlightenment thought through the prism of Priestley." —*Seattle Weekly*

"In *The Invention of Air* Steven Johnson gives a biography not just of a man, but a time in which the spigot of ideas was gradually being cranked wide open. It's a fun (and quite short) read for anyone interested in the intersection of science, politics, and religion. It's also an interesting look at how societies react—for good and ill—to periods of rapid change."
—Daily Kos

"A breath of fresh air . . . Johnson paints Priestley not as a man of the past but precisely the sort of figure the world needs more than ever: A searcher who shared his discoveries openly and willingly, crossed disciplinary boundaries with impunity and insight, who conceived of the world as a large laboratory . . . We live in troubling times, filled with signs of a great economic apocalypse, politicized science on topics from birth control to climate change and religious zealots who kill innocents rather than live peacefully with them. This is exactly the moment to learn from Priestley, who survived riots, threats of prosecution and other hardships and yet never doubted that 'the world was headed naturally toward an increase in liberty and understanding.'"
—*New York Post*

"Intelligent . . . Steven Johnson, who has a fine reputation for discerning trends and for his iconoclastic appreciation of popular culture, chooses his topics well. As a reminder of the underlying sanity and common sense of this country—a reminder perhaps much needed after the excesses of a displeasing presidential election campaign—*The Invention of Air* succeeds like a shot of the purest oxygen."
—Simon Winchester, *Publishers Weekly* (signature review)

"Arresting account of the career of Joseph Priestley . . . Johnson employs his customary digressiveness to great effect . . . Another rich, readable examination of the intersections where culture and science meet from a scrupulous historian who never offers easy answers to troubling, perhaps intractable questions."
—*Kirkus Reviews*

"Joseph Priestley (1733–1804) was a veritable Renaissance man, whose interests and skills ranged from science to religion to politics. Science writer Johnson (*The Ghost Map*) weaves together all of these themes and how they played out in his life, in early America, and among the Founding Fathers. He tells the story [of Priestley] in a reader-friendly manner that also encourages readers to think about how these themes apply in today's world."
—*Library Journal*

*continued . . .*

# Praise for *The Ghost Map*

"By turns a medical thriller, detective story, and paean to city life, Johnson's account of the outbreak and its modern implications is a true page-turner."
—*The Washington Post*

"Johnson adds [an] . . . old-fashioned storytelling flair, another form of street knowledge—to his fractal, multifaceted method of unraveling the scientific mysteries of everyday life."   —*Los Angeles Times Book Review*

"A formidable gathering of small facts and big ideas . . . There's a great story here, one of the signal episodes in the history of medical science, and Johnson recounts it well."   —*The New York Times Book Review*

"[An] unputdownable tale . . . an ambitious and compelling work . . . Mr. Johnson is never less than lively and beguiling."   —*The Wall Street Journal*

"[Johnson] has latched on to a truly compelling story, and he calls in the voices of Charles Dickens, Samuel Pepys, George Eliot, Jane Jacobs, and Stephen Jay Gould to help tell it. They make a lively chorus."
—*The Cleveland Plain Dealer*

"Brings to nightmarish, thought-provoking life a world in which a swift but very unpleasant death can be just a glass of water away."
—*Entertainment Weekly*

"Steven Johnson tells the tale with verve, spicing his narrative with scenes of Dickensian squalor and the vibrant street life surrounding that squalor. But in Johnson's hands, *Ghost Map* morphs into something more than mere history."   —*The San Diego Union-Tribune*

"An engrossing story that should appeal to anyone interested in the idea of cities as functioning systems—not to mention tales of urban lechery and bounties of choice Charles Dickens quotes."   —*The Onion AV Club*

"[A] tightly written page-turner . . . From Snow's discovery of patient zero to Johnson's compelling argument for and celebration of cities, this makes for an illuminating and satisfying read."   —*Publishers Weekly* (starred review)

## Praise for *Everything Bad Is Good for You*

"Wonderfully entertaining . . . Steven Johnson proposes that what is making us smarter is precisely what we thought was making us dumber: popular culture. . . . There is a pleasing eclecticism to [Johnson's] thinking. He is as happy analyzing *Finding Nemo* as he is dissecting the intricacies of a piece of software . . . Johnson wants to understand popular culture . . . in the very practical sense of wondering what watching something like *The Dukes of Hazzard* does to the way our minds work."

—Malcolm Gladwell, *The New Yorker*

"Iconoclastic and captivating . . . *Everything Bad Is Good for You* is a lucid tour of the pop-culture landscape."
—*The Boston Globe*

"Revelatory . . . Daring . . . Finally, an intellectual who doesn't think we're headed down the toilet!"
—*The Washington Post Book World*

"Persuasive . . . old dogs won't be able to rest as easily once they've read *Everything Bad Is Good for You*, Steven Johnson's elegant polemic. . . . It's almost impossible not to agree with him."

—Walter Kirn, *The New York Times Book Review*

"A thought-provoking argument that today's allegedly vacuous media are, well, thought-provoking . . . A brisk, witty read, well versed in the history of literature and bolstered with research . . . Johnson, it turns out, still knows the value of reading a book. And this one is indispensable."
—*Time*

"The author *Newsweek* called one of the most influential people in cyberspace . . . is back. The beauty of Johnson's latest work—beyond its engaging, accessible prose—is that anyone with even a glancing familiarity with pop culture will come to the book ready to challenge his premise. *Everything Bad Is Good for You* anticipates and refutes nearly every likely claim, building a convincing case that media have become more complex and thus make our minds work harder."
—*The Cleveland Plain Dealer*

"Very entertaining."
—*Time Out New York*

**RIVERHEAD BOOKS**

*New York*

# THE
# INVENTION
## OF AIR

A STORY OF **SCIENCE, FAITH,**

**REVOLUTION,** AND THE

**BIRTH OF AMERICA**

# STEVEN
# JOHNSON

**RIVERHEAD BOOKS**
Published by the Penguin Group
Penguin Group (USA)
375 Hudson Street, New York, New York 10014, USA

USA I Canada I UK I Ireland I Australia I New Zealand I India I South Africa I China

Penguin Books Ltd., Registered Offices: 80 Strand, London WC2R 0RL, England
For more information about the Penguin Group, visit penguin.com.

The Library of Congress has catalogued the Riverhead hardcover edition as follows:

Johnson, Steven, date.
The invention of air : a story of science, faith, revolution,
and the birth of America / Steven Johnson.
p.      cm.
Includes bibliographical references and index.
ISBN 978-1-59448-852-8
1. Priestley, Joseph, 1733–1804.   2. Chemists—Great Britain—Biography.
3. Scientists—Great Britain—Biography.   I. Title.
QD22.P8J635      2008           2008046101
540.92—dc22
[B]

First Riverhead hardcover edition: December 2008
First Riverhead trade paperback edition: October 2009
Riverhead trade paperback ISBN: 978-1-59448-401-8

PRINTED IN THE UNITED STATES OF AMERICA

15   14   13   12   11   10   9   8   7   6   5   4

Cover design © 2008 Keenan
Cover photo of Joseph Priestley: English School, The Bridgeman Art Library © Getty Images
Cover illustration of pneumatic bath: Hulton Archive/Stringer © 2007 Getty Images
Book design by Chris Welch

*For Jay*

*The English hierarchy (if there be anything unsound in its constitution) has equal reason to tremble at an air pump, or an electrical machine.* —JOSEPH PRIESTLEY

*That ideas should freely spread from one to another over the globe, for the moral and mutual instruction of man, and improvement of his condition, seems to have been peculiarly and benevolently designed by nature, when she made them, like fire, expansible over all space, without lessening their density at any point, and like the air in which we breathe, move, and have our physical being, incapable of confinement or exclusive appropriation.* —THOMAS JEFFERSON

# CONTENTS

# AUTHOR'S NOTE

A few days before I started writing this book, a leading candidate for the presidency of the United States was asked on national television whether he believed in the theory of evolution. He shrugged off the question with a dismissive jab of humor. "It's interesting that that question would even be asked of someone running for president," he said. "I'm not planning on writing the curriculum for an eighth-grade science book. I'm asking for the opportunity to be president of the United States."

It was a funny line, but the joke only worked in a specific intellectual context. For the statement to make sense, the speaker had to share one basic assumption with his audience: that "science" was some kind of specialized intellectual field, about which political leaders needn't know anything to do their business. Imagine a candidate dismissing a question about his foreign policy experience by saying he was running for president and not writing a textbook on international affairs. The joke wouldn't make sense, because we assume that foreign policy expertise is a central qualification for the chief executive. But science? That's for the guys in lab coats.

That line has stayed with me since, because the web of events at the center of this book suggests that its basic assumptions are fundamentally flawed. If there is an overarching moral to this story, it is that vital fields of intellectual achievement cannot be cordoned off from one another and relegated to the specialists, that politics can and should be usefully informed by the insights of science. The protagonists of this story lived in a climate where ideas flowed easily between the realms of politics, philosophy, religion, and science. The closest thing to a hero in this book—the chemist, theologian, and political theorist Joseph Priestley— spent his whole career in the space that connects those different fields. But the other figures central to this story—Ben Franklin, John Adams, Thomas Jefferson—suggest one additional reading of the "eighth-grade science" remark. It was anti-intellectual, to be sure, but it was something even more incendiary in the context of a presidential race. It was positively un-American.

In their legendary fourteen-year final correspondence, reflecting back on their collaborations and their feuds, Thomas Jefferson and John Adams wrote 165 letters to each other. In that corpus, Benjamin Franklin is mentioned by name five times, while George Washington is mentioned three times. Their mutual nemesis Alexander Hamilton warrants only two references. By contrast, Priestley, an Englishman who spent only the last decade of his life in the United States, is mentioned fifty-two times. That statistic alone gives some sense of how important Priestley was to the founders, in part because he would play a defining role in the rift and ultimate reconciliation between Jefferson and Adams, and in part because his distinctive worldview had a profound impact on both men, just as it had on Franklin three decades before. Yet today, Priestley is barely more than a footnote in most popular accounts of the revolutionary generation. This book is an attempt to understand how Priestley became so central to the great minds of this period—in the fledgling United States, but also in England and France. It is not so much a biography as it is the biography of one man's ideas, the links of association and influence that connect him to epic changes in science, belief, and society—as well as to some of the darkest episodes of mob violence and political repression in the history of Britain and the United States.

One of the things that makes the story of Priestley and his peers so fascinating to us now is that they were active participants in revolutions in multiple fields: in politics, chemistry, physics, education, and religion. And so part of my intent with this book is to grapple with the question of why these revolutions happen when they do, and why some rare individuals end up having a hand in many of them simultaneously. My assumption is that this question cannot be answered on a single scale of experience, that a purely biographical approach, centered on the individual life of the Great Man and his fellow travelers, will not do it justice; nor will a collectivist account that explains intellectual change in terms of broad social movements. My approach, instead, is to cross multiple scales and disciplines—just as Priestley and his fellow travelers did in their own careers. So this is a history book about the Enlightenment and the American Revolution that travels from the carbon cycle of the planet itself, to the chemistry of gunpowder, to the emergence of the coffeehouse in European culture, to the emotional dynamics of two friends compelled by history to betray each other. To answer the question of why some ideas change the world, you have to borrow tools from chemistry, social history, media theory, ecosystem science, geology. That connective sensibility runs against the grain of our specialized intellectual culture, but it would have been second nature to Priestley, Franklin, Jefferson, Adams, and their peers. Those are our roots. This book is an attempt to return to them.

# THE
# INVENTION
## OF AIR

Joseph Priestley

# The Vortex

*May 1794*
*The North Atlantic*

THE FIRST SIGN OF A WATERSPOUT FORM-
ing is a dark stain on the surface of the sea, like a circle of
black ink. Within a matter of minutes, if atmospheric conditions are right, a spiral of light and dark streaks begins to
spin around the circle. Soon a ring of spray rises up into the
air, water molecules propelled aloft by the accelerating winds
at its periphery. And then the spout surges to life, a whirling
line drawn from sea to sky, sustained by rotational winds that
have been measured at up to 150 miles per hour.

Unlike land-based tornados, waterspouts often form in
fair weather: a vortex of wind, capable of destroying small
vessels, that appears, literally, out of the blue. While it is
not nearly as dangerous as a traditional tornado, the waterspout was long a figure of fear and wonder in mariner tales
of life on the open sea. In the first century B.C., Lucretius

described "a kind of column [that] lets down from the sky into the sea, around which the waters boil, stirred up by the heavy blast of the winds, and if any ships are caught in that tumult, they are tossed about and come into great peril." Sailors would pour vinegar into the sea and pound on drums to frighten off the spirits that they imagined lurking in the spout. They had good reason to be mystified by these apparitions. The upward pull of the vortex is strong enough to suck fish, frogs, or jellyfish out of the water and carry them into the clouds, sometimes depositing them miles from their original location. Scientists now believe that apocryphal-sounding stories of fish and frogs raining from the sky were actually cases where waterspouts gulped up a menagerie of creatures straight out of the water, and then deposited them on the heads of bewildered humans when the spout crossed over onto land and dissipated.

A waterspout sighting is a meteorological rarity, even in the tropical waters where spouts are most often seen. Ships in the colder waters of the North Atlantic, particularly during early spring, almost never encounter them. So it was more than a little surprising that, on one extraordinary day in the spring of 1794, the hundred-odd passengers en route to New York aboard the merchant ship *Samson* caught sight of four distinct waterspouts simultaneously drifting their way across the sea.

Most passengers onboard the *Samson* would have viewed the looming spouts not as statistical anomalies but as sinister omens, if not outright threats. No doubt some passengers aboard the *Samson* ran belowdecks in fear at the first sighting, while others stared in wonder at the four spouts. But we can say with some confidence that one passenger aboard the *Samson* rushed to the deck at the first hint of a waterspout sighting, and stood transfixed, observing the spray patterns and cloud formations. It is easy to imagine him borrowing the captain's telescope and peering into the vortex, estimating wind velocity, perhaps jotting down notes as he watched. He would have known that the lively scientific debate over spouts—started in part by his old friend Benjamin Franklin—revolved around whether spouts descended from clouds, as tornados do, or whether they propelled themselves upward from the ocean surface. The idea of witnessing four waterspouts on a North Atlantic voyage would not have been a sign of foreboding or an imminent threat for him. It would have been a stroke of extraordinary good luck.

This was Joseph Priestley, formerly of Hackney, England, en route to his new home in America. At sixty-one years old, he was among the most accomplished men of his generation, rivaled only by Franklin in the diversity of his interests and influence. He had won the Copley Medal (the Nobel Prize of its day) for his experiments on various gases in his late thirties, and published close to five hundred books and pamphlets on science, politics, and religion since 1761. An ordained minister, he had helped found the dissenting Christian sect of Unitarianism. He counted among his closest friends the great minds of the Enlightenment and the early Industrial Revolution: Franklin, Richard Price, Josiah Wedgwood, Matthew Boulton, James Watt, Erasmus Darwin.

But while Priestley's luminous career had established an extensive base of admirers in the newly formed United States, he had booked passage on the *Samson* thanks to another, more dubious, honor. He had become the most hated man in all of Britain.

TRANSATLANTIC VOYAGES in the late eighteenth century were perilous affairs, even when the vessel avoided the substantial risk of being "lost at sea." One of the most ghastly accounts of sea travel from that period—Gottlieb Mittelberger's *Journey to Pennsylvania*—described the scene onboard the ship *Osgood* as it made its way from Rotterdam to Philadelphia in the summer of 1750:

> But during the voyage there is on board these ships terrible misery, stench, fumes, horror, vomiting, many kinds of sea-sickness, fever, dysentery, headache, heat, constipation, boils, scurvy, cancer, mouth-rot, and the like. . . . Add to this want of provisions, hunger, thirst, frost, heat, dampness, anxiety, want, afflictions and lamentations, together with other trouble, as . . . the lice abound so frightfully, especially on sick people, that they can be scraped off the body. . . . The water which is served out on the ships is often very black, thick and full of worms, so that one cannot drink it without loathing, even with the greatest thirst. . . . Towards the end [of the *Osgood*'s voyage] we were compelled to eat the ship's bis-

cuit which had been spoiled long ago, though in a whole biscuit there was scarcely a piece the size of a dollar that had not been full of red worms and spider's nests.

It was not exactly the *Queen Mary*, to say the least. A nice clean shipwreck might have started to seem appealing after a few days dining on black wormwater and spider's eggs. On the *Samson*, the drunken captain and his first mate argued so violently with each other that the water casks were neglected and caused much "suffering" among the steerage passengers, according to Priestley's somewhat ambiguous account. Mary Priestley, Joseph's wife, labored through three weeks of constant seasickness in the heavy seas that the *Samson* met upon leaving England.

To embark on such a journey at the age of sixty-one took a particular mix of fearlessness and optimism. Priestley had both qualities in abundance. Nearly every extended description of the man eventually winds its way to some comment about his relentlessly sunny outlook. He was almost pathologically incapable of believing the threats that arrayed themselves against him. Here is Priestley giving his account of the voyage of the *Samson*, in a letter written to a friend upon landing in New York:

> We had many things to amuse us on the passage; as the sight of some fine mountains of ice; water-spouts, which [are] very uncommon in those seas; flying fishes, porpoises, whales, and sharks, of which we caught one; luminous sea-water, &c.

The storm that nearly sunk the ship merits two brief sentences, amid all the amusements:

> We had very stormy weather, and one gust of wind as sudden and violent as, perhaps, was ever known. If it had not been for the passengers, many of the sails had been lost.

Mary Priestley was less sanguine about the storm ("It was a very awful night") and struggled to strike a similar note of enthusiasm in her description of the passing diversions of the voyage:

> Our voyage at times was very unpleasant, from the roughness of the weather; but as variety is charming, we had all that could well be experienced on board, but shipwreck and famine.

It's not hard to hear a hint of gritted teeth or gentle satire in that "variety is charming" line, as though she's mimicking a discourse from her beloved "Dr. P" on the latest sighting of "luminous sea-water" or some other fascination—a speech she had heard a few too many times during those three weeks of seasickness.

But however severe the peril that confronted them in setting sail for America, in that spring of 1794, Mary and Joseph Priestley had little choice but to book passage on the *Samson*. The open rage and violence that had rained down on them made the decision to flee inevitable. Priestley had spent weeks shuttling from safe house to safe house, as the newspapers and pamphleteers and cartoonists called for his head. His persecution had caused many to compare him to Socrates. (Before Priestley's departure, then vice president John Adams wrote in a letter to Priestley, "Inquisitions and Despotisms are not alone in persecuting Philosophers. The people themselves, we see, are capable of persecuting a Priestley, as another people formerly persecuted a Socrates.") In contemporary terms, Priestley had become the Salman Rushdie of Georgian England: a world-famous intellectual whose political and theological musings had planted a bull's-eye on his back. America was the logical way out.

DURING THE CALM DAYS on the second half of the Samson's voyage, Priestley would stand at the stern of the ship and lower a thermometer attached to a rope into the sea to record the temperature of the water at different depths. Such exact measurements would have been impossible at the beginning of the century; the sealed mercury thermometer had been invented in 1714 by Daniel Gabriel Fahrenheit, who also devised a scale for his contraption, establishing 32 degrees as the freezing point. As is so often the case in the history of science, an increase in the accuracy of measurement led to a fundamental shift in the perception of the world. Marking changes in the temperature of ocean water enabled naviga-tors to identify and exploit a pattern in the ocean's currents that they had blindly stumbled across in centuries past: a river of warm water that runs from the tropics all the way up the coastline of North America, and then makes a sharp right turn toward Europe as it passes Cape Cod. Sailors had long tapped the energy of that oceanic river in their travels along the eastern seaboard, but its continued passage across the North Atlantic had gone largely undetected by all but the most experienced seamen.

The first precise measurement of that oceanic flow came indirectly through a pattern detected in the flow of information. In 1769, the Customs Board in Boston made a formal complaint to the British Treasury about the speed of letters arriving from England. (Indeed, regular transatlantic correspondents had long noticed that letters posted from America to Europe tended to arrive more promptly than letters sent the other direction.) As luck would have it, the deputy postmaster general for North America was in London when the complaint arrived—and so the British authorities brought the issue to his attention, in the hope that he might have an explanation for the lag. They were lucky in another respect: the postmaster in question happened to be Benjamin Franklin.

Franklin would ultimately turn that postal mystery into one of the great scientific breakthroughs of his career: a turning point in our visualization of the macro patterns formed by ocean currents. Franklin was well prepared for the task. As a twenty-year-old, traveling back from his first voyage to London in 1726, he had recorded notes in his journal about the strange prevalence of "gulph weed" in the waters of the

North Atlantic. In a letter written twenty years later, he had remarked on the slower passage westward across the Atlantic, though at the time he supposed it was attributable to the rotation of the Earth. In a 1762 letter he alluded to the way "the waters mov'd away from the North American Coast towards the coasts of Spain and Africa, whence they get again into the Power of the Trade Winds, and continue the Circulation." He called that flow the "gulph stream."

When the British Treasury came to him with the complaint about the unreliable mail delivery schedules, Franklin was quick to suspect that the "gulph stream" would prove to be the culprit. He consulted with a seasoned New England mariner, Timothy Folger, and together they prepared a map of the Gulf Stream's entire path, hoping that "such Chart and directions may be of use to our Packets in Shortning their Voyages." The Folger/Franklin map was the first known chart to show the full trajectory of the Gulf Stream across the Atlantic. But the map was based on anecdotal evidence, mostly drawn from the experience of New England—based whalers. And so in his voyage from England back to America in 1775, Franklin took detailed measurements of water temperatures along the way, and detected a wide but shallow river of warm water, often carrying those telltale weeds from tropical regions. "I find that it is always warmer than the sea on each side of it, and that it does not sparkle in the night," he wrote. In 1785, at the ripe old age of seventy-nine, he sent a long paper that included his data and the Folger map to the French scientist Alphonsus le Roy. Franklin's paper on "sundry Maritime Observations," as he modestly called it, delivered the first empirical proof of the Gulf Stream's existence.

So as Joseph Priestley dipped his thermometer into the waters of the Atlantic, he was retracing the steps that Franklin had taken almost twenty years before. The sight of those four waterspouts would also have brought back fond memories of his old friend. In his letter to le Roy, Franklin had speculated that North Atlantic waterspouts likely arose out of the collision between cold air and the warm water of the Gulf Stream. There is no direct evidence in the historical record, but it is entirely probable that it was the waterspout sighting that sent Priestley off on his quest to measure the temperature of the sea, trying to marshal supporting evidence for a passing conjecture his friend had made a decade before. Franklin had been dead for nearly four years, but their intellectual collaboration continued, undeterred by war, distance, even death.

Priestley's retracing of Franklin's 1775 journey went far beyond the scientific experiments they each performed en route. Franklin, too, had been a hunted man in his final days in London, driven from England by scandal and the first stirrings of war. Twenty years later, Priestley was making the same voyage, facing the same threat. While their religious beliefs differed, their scientific and political views were remarkably harmonious. In his intellectual sensibility, Franklin was closer to Priestley than he was to any of the American founding fathers. This was the bleak irony of their

parallel voyages across the Atlantic: the ideal of Enlightenment science had instilled in them a set of shared political
values, a belief that reason would ultimately triumph over
fanaticism and frenzy. But now the vortex had swallowed
them both.

All around Priestley immense forces of energy surged:
the tight spiral of the waterspout, the vast conveyer belt of
the Gulf Stream, the liberated energy of the British coal
fields that had helped send him into exile. One of Priestley's
greatest scientific discoveries involved the cycle of energy
flowing through all life on Earth, the origin of the very air he
was breathing there on the deck as he watched his thermometer line bob in the waters of the Atlantic. Together, all those
forces converged on him, as the *Samson* struggled against
the current, bearing west to the New World . . .

BENJAMIN FRANKLIN AND THE KITE

## CHAPTER ONE

# The Electricians

*December 1765*
*London*

THE LONDON COFFEE HOUSE LAY IN ST.
Paul's churchyard, a crowded urban space steps from the
cathedral, bustling with divinity students, booksellers, and
instrument makers. The proximity to the divine hadn't
stopped the coffeehouse from becoming a gathering place
for some of London's most celebrated heretics, who may
well have been drawn to the location for the sheer thrill of
exploring the limits of religious orthodoxy within shouting
distance of England's most formidable shrine. On alternat-
ing Thursdays, a gang of freethinkers—eventually dubbed
"The Club of Honest Whigs" by one of its founding members, Benjamin Franklin—met at the coffeehouse, embark-
ing each fortnight on a long, rambling session that has no
exact equivalent in modern scientific culture. (The late-night
bender at an industry conference probably comes closest: the

sharing of essential, potentially lucrative information while stimulated by the chemical cocktail of caffeine, alcohol, and nicotine.) Boswell visited the "Honest Whigs" on occasion, and he had this to say of the experience:

> It consists of clergymen, physicians and some other professions . . . (including) Mr Price who writes on morals . . . we have wine and punch upon the table. Some of us smoke a pipe, conversation goes on pretty formally, sometimes sensibly and sometimes furiously: At nine there is a sideboard with Welsh rabbits and apple-puffs, porter and beer.

On December 19, 1765, Joseph Priestley sat down at a coffeehouse table, there in the shadow of St. Paul's, and began a conversation that would transform his life. London had dazzling sights, and shops full of the latest scientific equipment, and Royal Societies devoted to pioneering research. But like so many young men and women since, Priestley had come to the great city with one driving objective: he had a book idea to pitch. That was why he found himself, for the first time, in the good company of the Honest Whigs.

Priestley was thirty-two, an affable and freethinking minister and schoolteacher whose career to date had been somewhat stymied by a persistent stammer. (His first trip to London, ten years earlier, had been to spend a month with a speech therapist, a Mr. Angier, who promised to "cure all defects of speech" and made his clients take an oath not to reveal his technique.) Born in 1733 in a small town called Field-head, about six miles outside of Leeds, Priestley belonged to an extended family of religious nonconformists, at a time of intense political and theological battles between the Church of England and religious dissenters. Even in that unorthodox milieu, Priestley managed to push the boundaries: at nine-teen, he was denied membership in the Independent Cha-pel of Heckmondwike, in Yorkshire. Exiled from the strict Calvinism of his family, he spent his twenties preaching to small dissenting congregations in Needham and Nantwich, offending a few parishioners along the way with his maver-ick theories on the divinity of Jesus Christ.

The congregation in Nantwich numbered only sixty regu-lar attendees, which left Priestley with plenty of spare time to start a small school in the town, where he instructed thirty boys six days a week. He began writing in that period, drafting short treatises on theological matters—the supernatural dis-tortions of the Apostle Paul was a favorite subject—showing them to a few mentors and then burying them in his drawer for later revision. And while those essays would eventually find their way to a mass readership, his first published book was an equally radical take on a seemingly less contentious subject, *The Rudiments of English Grammar*, one of the first attempts to systematically map the structure of the English language with the rigor that scholars had long applied to Latin and Greek. (Priestley's combination of innovative linguistics scholarship and firebrand political writing would chart a path followed two centuries later by Noam Chomsky.)

*Rudiments* helped Priestley land the post of tutor at Warrington Academy, a prominent dissenting school in Yorkshire. Originally hired to teach languages (he was fluent in six), Priestley quickly introduced courses in modern history and politics—a cutting-edge curriculum in an educational regimen still devoted to conjugating the verbs of dead languages. His first year at Warrington, Priestley wed Mary Wilkinson in Wrexham, Wales, where Mary's industrialist father ran the Bersham Ironworks. In his memoirs, Priestley would later write of his marriage: "This proved a very suitable and happy connexion, my wife being a woman of an excellent understanding, much improved by reading, of great fortitude and strength of mind, and of temper in the higher degree affectionate and generous; feeling strongly for others, and little for herself."

During his years at Nantwich, Priestley had developed an amateur's passion for science. Though barely able to make ends meet, by the late 1750s he had cobbled together enough savings to buy an air pump and an "electricity machine." Together with a well-calibrated scale, these three contraptions were at that time the state-of-the-art essentials of a scientific toolkit. (They would, each in their different ways, help support the great tower of scientific innovation that Priestley would build in the coming years.) By the time he got to Warrington, Priestley had the science bug. He had become, to use the terminology then in vogue, a dabbler in "natural philosophy."

Like many of his peers, his first love was electricity. To understand the importance of electricity in the imagination

of the educated classes in the mid-1700s, one has to understand the unusual convergence that made it so fascinating.
In most cases when a fundamental force in the universe is
first formally understood by science, there is a lag between
that understanding and the emergence of popular technologies that depend on the science for their existence. Newton's law of universal gravitation didn't immediately spawn
a craze for gadgets built on his equations. Even in today's
accelerated world, it took at least two generations for Watson
and Crick's discovery of DNA to engender mainstream technologies such as DNA tests. But with electricity, the two phenomena overlapped: you had the discovery of one of nature's
most fundamental forces, and you had an immediate flood
of mesmerizing parlor tricks. You had awe-inspiring scientific genius, and you had gadgets, all in one swoop.

Until the 1740s, electricity had been thought of as two
separate fluids, with the relationship between them poorly
understood. After conducting an ingenious run of experiments—many of which involved literally shocking his
houseguests with a machine designed to generate static electricity—Benjamin Franklin hit upon a series of fundamental
insights about electricity that remain unchallenged to this
day. Franklin first suggested that electricity was composed
of a single fluid, with two inseparable charges, which he
called "positive" and "negative." He discovered, likewise, that
the two charges interacted in predictable ways; the current
would reliably attempt to flow from a positively charged body
to a negatively charged one. From this, Franklin deduced the

general principle known as the "conservation of electrical charge"—the idea that electricity can neither be created nor destroyed, but instead is merely passed from one conducting object to another. (His biographer Walter Isaacson suggests that this insight may have originated in the many years Franklin spent poring over balance sheets as he built up his publishing business in Philadelphia.)

That basic model of electricity survives to this day, along with the vocabulary Franklin built to describe it. ("Battery," "charged," and "conductor" were all his coinages.) The gadgets, however, have not fared as well. Consider this drawing:

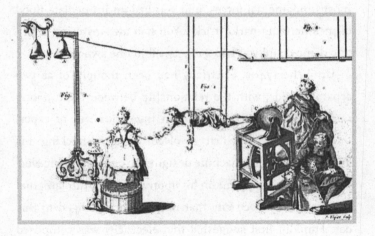

As bizarre as it looks, scenes like this were regular appearances in the drawing rooms and fairgrounds of the mid-eighteenth century. They were the special effects of Enlightenment popular culture. In this case, a young boy suspended in the air with silk ropes is positively charged by a machine that generates static electricity. First the boy's hair spikes up. Then, as the onlookers gasp in amazement, he reaches to touch a small girl, and sparks shoot between their fingertips. Willing volunteers were regularly pulled out of the audience to experience the voltage firsthand. The early explorers into this magical realm, scientists and showmen alike, were known by a name that also persists to the present day, though it has a somewhat different connotation now. They were called the Electricians.

The most transformative gadget to come out of the Electricians' cabinet of wonders was the lightning rod, also a concoction of Franklin's. (The quick jump from conceptual breakthrough to practical application was a hallmark of Franklin's science, as it would be of Priestley's.) Humans had long recognized that lightning had a propensity for striking the tallest landmarks in its vicinity, and so the exaggerated height of church steeples—not to mention their flammable wooden construction—presented a puzzling but undeniable reality: the Almighty seemed to have a perverse appetite for burning down the buildings erected in His honor.

Franklin first suggested the idea of taming that "electrical fire" in a letter to his friend Peter Collinson, written in 1750:

There is something however in the experiments of points, sending off, or drawing on, the electrical fire, which has not been fully explained, and which I intend to supply in my next. For the doctrine of points is very curious, and the effects of them truly wonderfull; and, from what I have observed on experiments, I am of opinion, that houses, ships, and even towns and churches may be effectually secured from the stroke of lightening by their means; for if, instead of the round balls of wood or metal, which are commonly placed on the tops of the weathercocks, vanes or spindles of churches, spires or masts, there should be put a rod of iron 8 or 10 feet in length, sharpen'd gradually to a point like a needle, and gilt to prevent rusting, or divided into a number of points, which would be better— the electrical fire would, I think be drawn out of a cloud silently, before it could come near enough to strike. . . .

Word of Franklin's hypothesis quickly spread, as his ideas circulated through the periodicals and coffeehouse networks, even crossing the Channel in a French translation. In 1752, the lightning-rod theory was first successfully put to the test (in France, as it turned out—the beginnings of Franklin's storied relationship with the French). Within five years of his speculative note to Collinson, lightning rods had become a common sight on church steeples throughout Europe and America. Franklin's biographer Carl Van Doren aptly describes the astonishment that greeted these events around the world: "A man in Philadelphia in America, bred a tradesman, remote from the learned world, had hit upon a secret which enabled him, and other men, to catch and tame the lightning, so dread that it was still mythological."

Thus it is no great surprise that when Joseph Priestley

took up the hobby of natural philosophy, it was electricity that first captured his fancy. No other field had generated so much scientific and practical innovation in such a short amount of time. But Priestley the writer had detected a missing piece in the growing science of electricity: no one had written a popular account of these world-changing discoveries. And so he had set off to London, hoping to meet the Electricians in the flesh, and to persuade them to let him tell the story of their genius.

PRIESTLEY ARRIVED in London armed with a letter of introduction from John Seddon, the rector at Warrington Academy, addressed to John Canton, a member of the Royal Society and a leading electrician himself. "You will find [Priestley] a benevolent, sensible man, with a considerable share of Learning," Seddon wrote. He added a postscript: "If Dr. Franklin be in Town, I believe Dr. Priestley would be glad to be made known to him."

Dr. Franklin did, in fact, prove to be in town, and so when Canton brought Priestley to the London Coffee House, the young, stammering schoolteacher from Warrington found himself seated across the table from the world's most celebrated electrician. They were joined by the Welsh moral philosopher and mathematician Richard Price, who would become one of Priestley's great friends and allies in the coming years.

The Honest Whigs had evolved out of a core group of

Canton's friends; most of them had been educated at Scottish universities or the dissenting schools that had cultivated Priestley. They were all, to varying degrees, convinced of the need for a "rational Christianity," though Franklin himself was said to abstain from most of the theological debates. Their politics were libertarian, and heated political debate often accompanied the "Welsh rabbits and apple-puffs." Boswell dryly relates one typical exchange: "Much was said this night against the parliament. I said that, as it seemed to be agreed that all Members of Parliament became corrupted, it was better to chuse men already bad, and so save good men." But the social and physical sciences often trumped politics at the coffeehouse: Price's breakthrough works on probability and demography (which would later influence Malthus) were rehearsed over wine and punch with the Honest Whigs. With so many prominent electricians in attendance, the conversations would invariably turn to the single-fluid theory, or a new hypothesis about conduction. A note survives in the historical record, sent from Franklin to Canton, making plans to travel together to the club, and asking, somewhat mysteriously, for "a little of his preparation for the Electrical Cushion."

Priestley had spent his entire life in small towns. Literally and figuratively, he lived on the periphery of the intellectual networks that consolidated in the metropolis. Given that background and his growing interests, it is easy to understand why he would have sought out an audience at the London Coffee House. These were his heroes, after all.

Despite their intimidating scholarship and cosmopolitan ways, the coffeehouse group was quick to embrace Priestley. He was personally likable, with a striking mix of intellectual acuity and gentleness. At five foot eight, he was tall for his era. (European men in the eighteenth century were more than two inches shorter on average.) Portraits of him from the period show a welcoming face, with sparkling gray eyes framed by a full-bottom wig. He was not as ruggedly handsome as some later hagiographic portraits would have it. But new acquaintances took to him immediately. While it is unclear exactly how much practical experimentation Priestley had done by 1765, there is little doubt that he possessed a firm understanding of the fledgling science of electricity. Speaking the lingua franca of the electricians would alone have probably warranted a warm greeting.

But the men had an even stronger reason to embrace the young minister: he had arrived on their doorstep offering to write a book in celebration of their research. With the hindsight of two centuries, Priestley's central idea seems an obvious one. A hundred compelling ideas and applications had spun out of the study of electricity in the past few decades. Wouldn't it be interesting for someone to string together the extended story of all those innovations in a single book? And do so in a way that made the tale intelligible to readers who lacked any specialized expertise?

Books about "experimental" or "natural" philosophy were not new, of course. Newton's *Philosophiæ Naturalis Principia Mathematica* had almost instantly revolutionized science when it appeared a century before. As the historian of science Thomas Kuhn writes, "No other work known to the history of science has simultaneously permitted so large an increase in both the scope and precision of research." The *Principia* even sold relatively well—Newton and his publisher, Edmund Halley, actually turned a small profit from it, despite its daunting content. But Newton had played by a set of genre conventions that limited the scope of his readership. Like other experimental philosophers of the age, Newton generally adopted a synthetic approach, one that, in the words of the historian Simon Shaffer, "presented discovery as a set of logically inevitable moves, and the achievement of discovery as an heroic act." The structure of the book was that of logical argument, the building of suppositions and proofs and counterarguments, all leading to Newton's own brilliant conceptual leap. The book form provided a model, of sorts, for the wider system he claimed to have uncovered in the physical world. The text orbited around his own genius.

Priestley had come to London with a vision of a different kind of book. He had seen more clearly than anyone of his era the possibility of science as a *narrative* experience. Newton had written a dazzling and inspired brief for his view of how the world worked. But it was ultimately his interpretation of the world that mattered, not the succession of earlier interpretations that had led the way to universal gravitation, despite his protestations about standing "on the shoulders of giants." Priestley saw the value in tracing a chain of events, turning it into a narrative of scientific progress. Newton wanted to persuade his readers to believe in a formula. Priestley wanted to tell them a story.

Newton also wrote in Latin, like almost all scholars of the period. Priestley's idea was to write his history in English, to ensure the widest possible reception. That popular touch would have particularly appealed to Franklin, who had built his career and public persona by publishing practical and lively essays for the eighteenth-century equivalent of the modern mass audience, and who had never bothered to embellish his scientific experiments with scholarly affectation.

When the evening at the London Coffee House finally came to an end, Joseph Priestley walked out into the churchyard with a new band of intellectual comrades and a promise of support for the intriguing book idea that he had outlined over the porter and wine. The electricians would open their private libraries and correspondence to him. (Simply tracking down the data had been the single biggest stumbling block to Priestley's history, given that public libraries and bookstores—not to mention Google—hadn't taken their modern form yet.) They promised to read the book in manuscript, and to suggest additions or corrections where appropriate.

Franklin, Canton, and Price took one other crucial step in their support of young Priestley: they encouraged him to conduct his own experiments while writing his history. With his *Rudiments of English Grammar* and his pamphlets, Priestley was already well on his way to a successful writing career when he first stepped foot into the coffeehouse. But hearing his idols urging him to write about his own investigations opened up a whole new field of possibility for the young man. A few days after that first meeting, the electricians took Priestley along to a session of the Royal Society, the apogee of English natural philosophy, where Newton himself had been president sixty years before. How thrilling it must have been for Priestley to walk into that sacred space with such illustrious new friends at his side. It was a story straight out of a nineteenth-century *Bildungsroman*, something from Balzac or Stendhal: a young man comes to the metropolis with big dreams and makes a name for himself. Priestley had arrived in London as a dabbler in natural philosophy, tinkering in the provinces with his electrical machine and his air pump. By the time he left, he was a scientist.

A FEW WEEKS LATER, after his return to Warrington, Priestley wrote to Canton: "The time I had the happiness to spend in your company appears in review like a pleasing dream. . . . I ardently wish a repetition of it." He spent the next year in a feverish rush, poring through the books and letters and pamphlets that his London friends had lent him, reconstructing the history of batteries, charges, lightning rods, and electrical fluids. He launched himself into a rapid and turbulent river of experiments, developing a style of investigation that would shape the rest of his career—more exploratory than systematic, shuffling through countless variations of materials and equipment and test subjects. Priestley was never one for the grand hypothesis; he rarely designed experiments specifically to test a general theory. The closest thing to a general theory in his work would ultimately lead to his greatest intellectual mistake. His approach was far more inventive, even chaotic. While the experiments themselves were artfully designed, his higher-level plan for working through a sequence of experiments was less rigorous. Priestley's mode was to get interested in a problem—conductivity, fire, air—and throw the kitchen sink at it. (Literally so, in that many of his experiments were conducted in a kitchen sink.) The method was closer to that of natural selection than abstract reasoning: new ideas came out of new juxtapositions, randomness, diversity. Priestley would later credit the emerging technology of the period—air pumps and electrostatic machines—with helping him develop his distinctive approach: "By the help of these machines," he wrote, "we are able to put an endless variety of things into an endless variety of situations, while nature herself is the agent that shows the result."

There is an almost comic quality to the incessant letters that Priestley sent his electrician friends in London over the spring and summer of 1766, postcards from the laboratory of a mad scientist:

I have made an experiment which, I think proves that *Glass when heated red hot is a conductor of electricity*. I took a glass tube about four feet long, and by means of mercury on the inside and tinfoil on the outside, I charged about nine inches of it very strongly. . . .

I took a cork, and stuck into the sides of it (pointing directly from the center) thirteen vanes each consisting of half a common card. Into the middle of the card I stuck a needle. . . .

I have made a great number of *experiments on animals,* some of which I refer to a letter I lately wrote to Dr Watson. Since I wrote to him, I discharged 37 Square feet of coated glass through the head and tail of a CAT three or four years old. She was instantly seized with universal convulsions, then lay as dead a few seconds. . . . Thinking she would probably die a lingering death in consequence of the stroke, I gave her a second, about half an hour after the first. She was seized as before, with universal convulsions, and in the convulsive respiration which succeeded she expired. She was dissected with great care, but nothing particular was observed.

Early in his 1766 investigations, Priestley thought he had stumbled across a crucial observation: "mephitic" air—now known as carbon dioxide—was a conductor of electricity. He wrote excitedly to Canton with the news, only to discover in the coming weeks that the results had been compromised by small molecules of condensed water in the glass that held the air. (Water was already a well-known conductor.) He wrote a sheepish letter to his electrician friends retracting his earlier claims, but the experiment ultimately led him to one of his most important contributions to the science of electricity: the addition of charcoal to the then short list of substances that were capable of conduction, alongside water and metal.

By the end of 1766, a more fundamental pattern had emerged out of the chaos of Priestley's electrical investigations. Building on a puzzling experiment that Franklin had devised using an "electrical cup," Priestley surmised that the relationship between electrical charges followed the same inverse square law that Newton had observed in gravitational attraction. (In layman's terms, the idea was that as two charges approached each other, the electrostatic force between them increased dramatically.) Two decades later, the French physicist Charles-Augustin de Coulomb would definitively prove that Priestley's conjectures were accurate, which is why the equation now goes by the name of Coulomb's Law, though Priestley was the first to propose it. It remains one of the bedrock principles of physics. Coulomb's Law would ultimately be deployed to explain why atoms attach to each other in forming molecules—why the world is made up of *stuff*, rather than diffuse gases. It would also play a central role in the invention of semiconductors and integrated circuits, the core technology that created the electronic and digital revolutions of the late twentieth century.

The constant flow of letters to London documenting his progress had impressed Priestley's electrician friends so much that by June, Messrs. Price, Franklin, and Canton decided to nominate their ambitious friend from Warrington as a member of the Royal Society:

Joseph Priestley of Warrington, Doctor of Laws, author of a chart of Biography, & several other valuable works, a gentleman of great merit & learning, & very well versed in

Mathematical & philosophical enquiries, being desirous of offering himself as a candidate for election into this Society, is recommended by us on our personal knowledge, as highly deserving that honour; & we believe that he will, if elected, be a usefull & valuable member.

As the year progressed, Priestley's letters were increasingly accompanied by chapters (Priestley called them "numbers," in the parlance of the day) from his growing manuscript. Somehow in the stretch of about fifteen months, Priestley had managed to write seven hundred pages on electricity and its pioneers, while exploring an "endless variety of situations" with his own experiments.

When *The History and Present State of Electricity, with Original Experiments* was published in 1767, the book instantly landed Priestley in that upper echelon of electricians that had welcomed him so warmly at the London Coffee House. A forty-page review in the *Monthly Review* called it "excellent . . . judicious, and well-informed." It sold well enough to support five English editions, and was subsequently translated into both French and German. Copies circulated around the globe: the Italian electrician Alessandro Volta read it; Franklin sent multiple copies back to the colonies. (By 1788, it was part of the standard natural philosophy curriculum at Yale.) The book would remain the principal text on electricity for nearly a hundred years.

The *History* began with a stirring argument for why electricity was so interesting in the first place:

Hitherto philosophy has been chiefly conversant about the more sensible properties of bodies; electricity, together with chemistry, and the doctrine of light and colours, seems to be giving us an inlet into their internal structure, on which all their sensible properties depend. By pursuing this new light, therefore, the bounds of natural science may possibly be extended, beyond what we can now form an idea of. New worlds may open to our view, and the glory of the great Sir Isaac Newton himself, and all his contemporaries, be eclipsed, by a new set of philosophers, in quite a new field of speculation. Could that great man revisit the earth, and view the experiments of the present race of electricians, he would be no less amazed than Roger Bacon, or Sir Francis, would have been at his.

Priestley condensed all of his own discoveries into the closing two hundred pages of the book, leaving the first five hundred to an exhaustive narrative of scientific progress, relating each innovation in careful detail.

He even included a few sections in the middle of the book that offered guidance for the aspiring scientists and show-men in his audience: "Practical maxims for the use of young electricians" and "A description of the most ENTERTAINING EXPERIMENTS performed by electricity." Those sections may not sound all that scholarly to the modern ear, but they were crucial to the underlying objectives of the book. Priestley aimed to popularize not simply by helping ordinary readers under-stand the new science of electricity, but also by encouraging them to become scientists themselves. While he wanted to celebrate the electricians' discoveries, he deliberately avoided establishing an aura of otherworldly genius around them:

> Were it possible to trace the succession of ideas in the mind of Sir Isaac Newton, during the time he made his greatest discoveries, I make no doubt but our amazement at the extent of his genius would a little subside. . . . [T]he interests of science have suffered by the excessive admiration and wonder with which several first rate philosophers are considered; and . . . an opinion of the greater equality of mankind in point of genius would be of real service in the present age.

The *History* was a seminal achievement in Enlightenment science for two distinct reasons. First, there were Priestley's original contributions to the science, the ideas that had won him admiration in London and landed him in the Royal Society. (In the style of Newton, Priestley had also included a number of unanswered questions and potential avenues for exploration that his successors would fruitfully investigate in the coming years.) But his *History* was as much a breakthrough for its form as for its content. He had invented a whole new way of *imagining* science; instead of a unified, Newtonian pronouncement, Priestley recast natural philosophy as a story of progress, a rising staircase of enlightenment, with each new innovation building on the last. In his prologue to the *History*, he contrasts his method favorably with the existing genres of civil history and natural history: the epic stories of kings and wars and famines, or the meticulous inventories of nature—insects, rock formations, flowers—that had become commonplace over the preceding century. There were great lessons and pleasures to be found in those other forms of writing, Priestley argued, but they lacked the definitive movement toward clarity and understanding that could be found in his own philosophical history:

The History of Electricity is a field full of pleasing objects, according to all the genuine and universal principles of taste, deduced from a knowledge of human nature. Scenes like these, in which we see a gradual rise and progress in things, always exhibit a pleasing spectacle to the human mind. . . . This pleasure, likewise, bears a considerable resemblance to that of the sublime, which is one of the most exquisite of all those that affect the human imagination. For an object in which we see a perpetual progress and improvement is, as it were, continually rising in its magnitude; and moreover, when we see an actual increase, in a long period of time past, we cannot help forming an idea of an unlimited increase in futurity; which is a prospect really boundless, and sublime.

In one sense, we can see Priestley inventing in these passages an entire genre of popular science: tales of discovery and exploration designed to captivate and engage the mind of a generalist reader. (Priestley would publish an even more accessible, and shorter, version of his *History* the following year, which he called *A Familiar Introduction to the Study of Electricity.*) But he is also building a specific way of connecting past and future that would animate his writing and thinking for the rest of his life, and that would profoundly shape the worldview of the American founders as well. Looking backward over the history of electricity enabled him both to appreciate how science had become an engine of progress and improvement and to project forward into the future, to imagine that ascending line, its trajectory continuing through the coming centuries. This is one of the origin points for a distinctly modern view of the world—call it progressive futurism. Countless other cultures had imagined themselves living at the apex of history and human understanding. Priestley took that assumption, grounded it in an empirical story of scientific discovery, and then added the crucial caveat: This is only the beginning!

> To look down from the eminence, and to see, and compare all those gradual advances in the ascent, cannot but give the greatest pleasure to those who are seated on the eminence, and who feel all the advantages of their elevated situation. And considering that we ourselves are, by no means, at the top of human science; that the mountain still ascends beyond our sight, and that we are, in fact, not much above the foot of it, a view of the manner in which the ascent has been made, cannot but animate us in our attempts to advance still higher, and suggest methods and expedients to assist us in our farther progress.

But even if eighteenth-century Europe was still miles away from the peak, Priestley nonetheless made it clear in his *History* which mountaineer had reached the highest elevation to date. He devoted almost a hundred pages to Ben Franklin's experiments and theories about electricity. On page 160 of the original printing, in a chapter devoted to Franklin's probing of the connection between lightning and electricity, Priestley launched into the story of a curious experiment that Franklin had devised in Philadelphia fifteen years before:

> To demonstrate, in the completest manner possible, the sameness of the electric fluid with the matter of lightning, Dr. Franklin, astonishing as it must have appeared, contrived actually to bring lightning from the heavens, by means of an electrical kite, which he raised when a storm of thunder was perceived to be coming on. . . . [S]o capital a discovery as this (the greatest, perhaps, that has been made in the whole compass of philosophy, since the time of Sir Isaac Newton) cannot but give pleasure to all my readers. . . .

The classic image of Franklin with his electrified kite, ingrained in the minds of countless American schoolchildren over the past two centuries, dates back to this paragraph from Priestley's *History*. Franklin himself had only published a brief third-person account of his experiment in the *Pennsylvania Gazette*, without specifying that he himself had performed it. In fact, Franklin would never provide a direct account of his kite-flying experiment in any of his own published works, leading some subsequent scholars to suspect that the whole episode was a fabrication. But he willingly gave Priestley extensive details on the event. ("Dreading the ridicule which too commonly attends unsuccessful attempts in science, [Franklin] communicated his intended experiment to no body but his son, who assisted him in raising the kite.") Priestley's story was engineered to do more than just popularize the bold, life-threatening scientific adventures of his new friend. It was also an attempt to give Franklin partial credit for independently proving that lightning was electrical in nature. Three French scientists, inspired by Franklin's experiments, had constructed an iron rod that successfully drew lightning from the sky in May of 1752. Priestley pointedly ends his account of Franklin's kite with a coda: "This happened in June 1752, a month after the electricians in France had verified the same theory, but before he had heard any thing that they had done."

So many elements from Franklin and Priestley's future—the folklore and popular mythology, the intellectual camaraderie, the world-changing ideas—are bound together as a first draft in the pages of the History. Franklin had helped Priestley become one of the great scientists of the age, and he supplied the source material that Priestley used to build his progressive vision of history, a model that would govern his thinking for the remainder of his days. Priestley had, in turn, created an iconic portrait of his mentor, and planted him in the Enlightenment pantheon alongside Isaac Newton. Franklin with his kite remains the defining image of the practical scientific ingenuity of the American founding fathers. And we have Joseph Priestley to thank for it.

THE SUCCESS OF THE *HISTORY* and the alliance with the Honest Whigs catapulted Priestley into a new realm of influence and recognition. But it was only a preview of coming attractions. Over the next eight years, he would go on an intellectual streak of legendary proportions, making *two* groundbreaking discoveries, each one the sort of achievement that on its own would warrant inclusion in the pantheon of Enlightenment science. He would publish multiple papers on his electrical research, inventing new apparatuses for the creation of electrical charge and recording the first known sighting of what we now call an "oscillatory discharge," which would eventually be crucial to the technology of radio and television. He would isolate and name ten distinct gases, now understood as some of the building blocks of Earth's atmosphere, sparking a revolution in chemistry. Along the way, he would write more than fifty books and pamphlets on politics, education, and faith.

And if that list doesn't seem impressive enough: he would also invent soda water.

Before we turn to the specifics of this extraordinary chapter in Priestley's life, we should first consider the interpretative problem it forces us to confront: not just the *what* of what happened, but the *why*. Intellectual historians have long wrestled with the strangeness of this kind of streak.

The thinker plods along, publishing erratically, making incremental progress, and then, suddenly—the floodgates open and a thousand interesting ideas seem to pour out. It's no mystery that there are geniuses in the world, who come into life with innate cognitive skills that are nurtured and provoked by cultural environments over time. It's not hard to understand that these people are smarter than the rest of us, and thus tend to come up with a disproportionate share of the Big Ideas. The mystery is why, every now and again, one of these people seems to get a hot hand.

One possibility is that the whole concept of the hot hand is an illusion, a trick of the mind that exploits our woeful skills at probability analysis. If you dispersed innovations randomly across a group of people, and placed them at random intervals as well, a few clusters would undoubtedly appear where an individual researcher would churn through a series of breakthrough ideas in a short amount of time. We're naturally inclined to see a hot hand here, some extra dose of inspiration that triggered the streak in the first place, but in fact the streak would just be an offshoot of that random distribution, no more magical than a repeated coin toss that every now and then turns up heads ten times in a row. Two famous studies of streaks in sports—a basketball study by Stanford psychologist Amos Tversky and a baseball version conducted by the Harvard Nobel laureate Ed Purcell—found that hot hands were a figment of our imagination: the fact that a player has just made a free throw makes him no more or less likely to sink the next one. Even the humiliating nadir of the Baltimore

Orioles' o-for-21 losing streak that began their 1988 season was securely within the range of expected outcomes, given the 200,000 major league games that have been played in the modern era. As Stephen Jay Gould put it, in an essay that widely popularized these studies: "Nothing ever happened in baseball above and beyond the frequency predicted by coin-tossing models. The longest runs of wins or losses are as long as they should be, and occur about as often as they ought to." The one exception, Gould went on to concede, was DiMaggio's fifty-six-game hitting streak, so far above the predicted range that, in Gould's words, it "ranks as pure heart."

The question for intellectual history is whether streaks of innovation are more like the Orioles' dismal start in '88 or more like DiMaggio in the summer of '41—a fantasy of misinterpreted probability or the sign of some special force at work, a "zone" that somehow lowers the barriers to discovery and understanding. One reason to suspect the latter is that, unlike free throws, ideas are clearly cumulative in nature; solving one problem often gives you a new set of conceptual tools that help you solve the next problem that presents itself. But with Priestley, the mystery is not just that he was able to hit upon so many important ideas in such a brief time frame, it's also that those ideas were scattered across so many different fields.

There is a parallel mystery here, one level up the chain. Human cultures have a long track record of collective hot streaks, where clusters of innovations seem to burst into flame after centuries of darkness. (We have names like "Renaissance" precisely to mark exactly how extreme the transformation is.)

Priestley was a key participant in one of these cultural-phase transitions, what was described self-consciously at the time, by Kant and others, as the Enlightenment, a term that embraces both the widening of political and religious possibility in eighteenth-century Europe and the extensive application of the scientific method to problems that had previously been shrouded in darkness. There were literally dozens of paradigm shifts in distinct fields during Priestley's lifetime, watershed moments of sudden progress where new rules and frameworks of understanding emerged. Priestley alone was a transformative figure in four of them: chemistry, electricity, politics, and faith. Each paradigm shift on its own has its own internally consistent narrative that describes its path, explaining how we came to understand something like the single-fluid theory: a litany of hunches, experiments, published papers, and popularizations. But what we don't have is a convincing theory about the system that connects all these local innovations, that causes them to self-organize into something so momentous that we have to dream up a name like the "Age of Enlightenment" to describe it. Beneath those innovations some deeper force seems to be operating, a kind of intellectual plate tectonics driving a thousand tremors on the surface. In Priestley's mountain metaphor, it's not so much that we are climbing the slope, but that the mountain itself is being pushed higher by the force of those immense but unseen land masses colliding. But what is that force exactly—and how can we measure it?

You can see in those opening passages from *The History and Present State of Electricity* that Priestley was acutely aware of this problem; the structure of the book itself was designed, in a sense, to present that long-term progressive movement with maximum emphasis. This is a sensibility that was largely absent in the Renaissance, despite the achievements of that period; the hill-town cultures of northern Italy still imagined historical change as Fortune's wheel: rising, falling, waxing, waning. Beginning with Descartes and Bacon, a feeling began to emerge in Western Europe that history was charting another trajectory—not an endless cycle of rise and fall, but instead a steady climb upward. Priestley's book was an attempt to take that hunch and turn it into history.

By the time of his death, the premise that society and science were riding a kind of permanent escalator, ascending the slope at ever-increasing speed, would be widely accepted, and the debate would turn to the nature of the engine that was driving that process. For much of the nineteenth century, the engine was dialectics—first in the abstract approach that Hegel took in his *Philosophy of History*, and then in the materialist rendition of Marx and Engels that famously turned Hegel "on his head." Social and intellectual history, in this view, advanced according to the fundamental laws of dialectical progress, thesis confronting antithesis, and generating some higher-order synthesis out of that collision. The existence of this force was, for generations of thinkers, as immutable and ubiquitous as gravity itself, and yet the concept has a strange mysticism to it—even in Marx's more grounded economic version. Its origins are as a philosophical method, a way of working through an argument to reach a more advanced understanding. It's easy to understand why an individual logician might use the dialectical method to construct a proof. But why should uncoordinated, collective behavior follow dialectical patterns? Cultural change needn't necessarily take that particular shape; it's more intuitive, in fact, to think that it would mimic the characteristic patterns of other systems: waves, for instance, or epidemics, or information networks.

What Marx did grasp, more clearly than any thinker before him, was that the proper interpretative scale for understanding change and progress is larger and deeper than that of the individual human life, yet at the same time is grounded in the material world. You couldn't attribute change exclusively to exceptional people, and you couldn't attribute it to some external and nebulous spirit, the way Hegel had done. There were great thinkers and leaders and visionaries, to be sure—Marx held Hegel up as one of them, to a fault probably—but that "great man" view of historical change exposed only a small slice of the full story, because the creation and spread of new ideas and new ways of living are shaped by forces both greater and smaller than individual humans. Marx identified three new primary macro processes that deserved to be included in the narrative: the class struggle, the evolution of capital itself, and technological innovations. They were all, for different reasons, enormously valuable contributions to the project of making sense of historical change. And they were all fundamentally correct, at least in their contention that class identity, capital, and technological acceleration would be prime movers in the coming centuries, and that each one had an independent life, outside the direct control of human decision-makers. Humans made the steam engine, but the steam engine ended up remaking humanity, in ways that the original inventors never anticipated.

The contemporary view of intellectual progress is dominated by one book: Thomas Kuhn's *The Structure of Scientific Revolutions*, published in 1962, from which the now conventional terms "paradigm" and "paradigm shift" originate. By some measures, Kuhn's book was the most cited text in the last quarter of the twentieth century, and it regularly ranks among the most influential books of the entire century. In *Revolutions*, Kuhn set out to dismantle the idea that scientific progress happens in a linear fashion, as a series of indisputable facts unearthed one after another, each breakthrough another definitive step toward absolute truth. (Kuhn calls this the "development-by-accumulation" model.) Instead, he explained, "normal" science works within an established paradigm: a set of rules and conventions that govern the definition of terms, the collection of data, and the boundaries of inquiry. But over time, anomalies appear inside the paradigm: data that can't be explained, questions that can't be answered using the tools of the existing model. At that point, certain adventurous researchers begin practicing what Kuhn called "revolutionary science," reaching outside the boundaries of the old paradigm, inventing new rules and conventions that eventually cause the old paradigm to collapse. The classic case study for the concept of a paradigm shift is the Copernican revolution in astronomy, but in actual fact, the first extended story that Kuhn tells in *The Structure of Scientific Revolutions* is the paradigm shift in chemistry that took place in the 1770s, led by the revolutionary science of Joseph Priestley.

While Kuhn's system placed the scientist squarely at the center of intellectual change, it made an essential break from the folklore of individual genius that Priestley had himself questioned two centuries before. Kuhn demonstrated convincingly that science was not a straightforward pursuit of universal truth, the genius suddenly discovering new facts about the world by sheer force of intellect. Instead, innovations in science came out of a complicated play among insight, empirical study, and the conventions of a given paradigm. The facts themselves were bounded, and in part created, by the *cultural* prescriptions of the current model. The trouble with Kuhn's system, however, came from its own, self-professed conceptual boundaries. "Aside from occasional brief asides," Kuhn explained in the preface, "I have said nothing about the role of technological advance or of external social, economic, and intellectual conditions in the development of the sciences." In Kuhn's analysis, change happens because anomalies appear *inside* the rules and expectations of normal science. *External* changes—in technology, society, politics—do not appear as factors in this schema. Revolutionary science happens inside the lab, isolated from the tumult of the external world. But what happens when a scientific paradigm shift coincides with comparable revolutions in the structure of human society or religious belief? Surely there are causal links that connect them, particularly when one man lies at the center of so many simultaneous revolutions.

Is there a better organizing principle, a better metaphor for making sense of conceptual revolutions like those that Priestley helped bring about? One might be a twentieth-century concept that neither Priestley nor Marx had available to them, and which was still a new idea for Thomas Kuhn in 1962: the ecosystem. Ecosystem theory has changed our view of the planet in countless ways, but as an intellectual model it has one defining characteristic: it is a "long zoom" science, one that jumps from scale to scale, and from discipline to discipline, to explain its object of study: from the microbiology of bacteria, to the cross-species flux of nutrient cycling, to the global patterns of weather systems, all the way out to the physics that explains how solar energy collides with the Earth's atmosphere. The following diagram shows what ecosystem science looks like in practice.

This is the Bretherton diagram, prepared by a committee of scholars associated with NASA in the mid-eighties. It attempts to show the main dynamics of global ecosystems theory, the multidisciplinary field that goes by the name Earth System Science. The diagram looks formidable to the untrained eye, but it looks even more formidable to the *trained* eye, because the trained eye sees in a flash how many distinct disciplines are yoked together in this densely interconnected system. Economists, microbiologists, atmospheric physicists, marine biologists, geologists, urban historians, chemists: these are intellectual clans that historically have

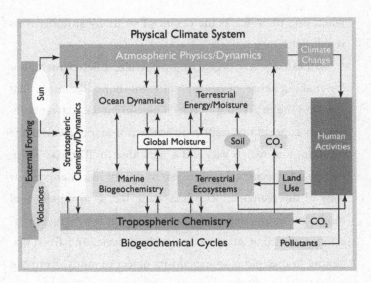

(FROM EARTH SYSTEM SCIENCE: AN OVERVIEW, NASA, 1988)

not spoken the same language, much less shared a table at the same coffeehouse. And yet there they are—connected, interdependent—on the Bretherton diagram. To make sense of the world system, they have had to learn to speak a common language.

Cultural systems, too—the natural history of good (and bad) ideas—require this kind of long thinking as well, from the neural networks of the human brain, to the biographical details of human lives, to the broad ebb and flow of social and physical energy in a changing society. The long zoom of culture looks something like this, moving from the very small to the very large:

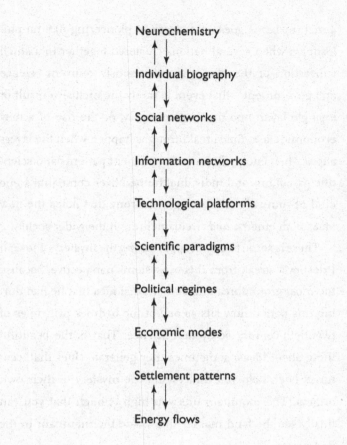

Neurochemistry

Individual biography

Social networks

Information networks

Technological platforms

Scientific paradigms

Political regimes

Economic modes

Settlement patterns

Energy flows

As in Earth System Science, each level operates at different time scales: biographical details of sibling rivalry or traumatic illnesses unfold on the scale of years or decades, while transformations in the flow of energy can take thousands (or millions) of years to play out. The economic base and the scientific paradigm figure prominently in this scheme, but neither has the primacy that Marx and Kuhn accorded to them. When something big happens in the culture—when a man in Leeds goes on a streak of pioneering natural philosophy; when several nations clustered together in a small subsection of the planet simultaneously reinvent science and government—that event is rarely the exclusive result of a single layer: one man's genius, say, or the rise of a new economic class. Epic breakthroughs happen when the layers align: when energy flows and settlement patterns and scientific paradigms and individual human lives come into some kind of mutually reinforcing synchrony that helps the new ideas both emerge *and* circulate through the wider society.

There is some poetry in approaching the mystery of Joseph Priestley's streak from this ecosystems perspective, because the most groundbreaking and original idea that he had during this period now sits as one of the bedrock principles of twentieth-century ecosystem science. That is the beautiful thing about ideas: sometimes they generate clues that, centuries later, help you understand the mystery of their own origins. The mountain lifts you high enough that you can finally see the land masses that made the mountain in the first place.

LONG-ZOOM HISTORIES don't dispense altogether with individual lives, of course, and in explaining Joseph Priestley's streak, it's best to start with one central biographical fact: he moved. In the summer of 1767, Joseph and Mary packed up their belongings at Warrington—the electrical kits and vials and growing library—so that Joseph could take up residence as minister to a congregation at Mill-Hill Chapel in Leeds. While the new job entailed preaching to a larger group of parishioners than in any of his previous positions, his daily obligations were far less imposing than they had been teaching at Warrington, requiring no more than an hour or two a day. With his wife running the household and tending to their four-year-old daughter, Sally, Priestley simply had more time on his hands to explore, invent, and write. Priestley was retracing a pattern that Franklin had originally carved two decades before, when he handed over day-to-day operation of his printing business to his foreman, David Hall, in 1748 and then spent the next three years transforming the science of electricity. Necessity may be the mother of invention, but most of the great inventors were blessed with something else: leisure time.

The move also inspired Priestley in more random ways. When the Priestleys first arrived in Leeds, they discovered the official minister's house on Bansinghall Street was still being renovated for them, and so they took up residence for a short while on Meadow Lane, in a house that happened to border on the public brewery of Jakes and Nell. Ever curious, Priestley quickly discovered that the vats of fermenting liquid emitted a steady supply of "fixed" or "mephitic" air—what we now call carbon dioxide. Fixed air had been discovered only a dozen years before by the Scottish chemist Joseph Black, who had been the first to propose that our atmosphere might in fact be a mixture of different elements, the poisonous "fixed" air intermingling with the common air that all animals require for respiration. Fixed air was almost as tantalizing a subject for inquiry as electricity in those days, and so within a matter of weeks, the puzzled workmen in the brewery were assisting the eccentric minister next door with a battery of experiments over the vats. Priestley discovered that pouring plain water back and forth between two cups while holding it over the vats suffused it with the fixed air after a short amount of time, adding an agreeable fizz that was reminiscent of certain rare mineral waters. In late September, he wrote a note to Canton describing his new fascination with mephitic air that included this aside: "By the way, I make most delightful *Pyrmont Water*, and can impregnate any water or wine &c. with that spirit in two minutes." If he had only thought to add fruit juice to the mix, he might have invented the wine cooler as well.

Priestley would refine his method in the coming years, and eventually mention his technique during a dinner party with the Duke of Northumberland in early 1772, suggesting—incorrectly as it turned out—that his seltzer water might prove a useful weapon in the British navy's fight against scurvy. Within a matter of days, Priestley was presenting a statement to the Lord Commissioners of the Admiralty on behalf of his concoction. By the time Captain Cook's vessels, the *Resolution* and the *Adventure*, set sail in June of 1772, they were equipped with soda-water machines manned by the watchful eye of the ships' surgeons. Inspired by the beverage's enthusiastic reception among the Admiralty, Priestley quickly published a pamphlet: *Directions for impregnating water with fixed air, in order to communicate to it the peculiar spirit and virtues of Pyrmont water, and other mineral waters of a similar nature.* Priestley's discovery did nothing to fight scurvy, but it did create a taste for carbonation that would ultimately conquer the planet.

Priestley later described his soda-water epiphany as his "happiest" discovery, while acknowledging it had little scientific value. But that chance encounter with the Jakes and Nell Brewery ultimately led to more substantive investigations as well: those fermenting vats with their invisible pool of mephitic air triggered in Priestley a new fascination with the mysteries of air itself, a fascination that would ultimately lead to the greatest discoveries of his career—along with his most vexing blunder. Had the renovations to the minister's house on Bansinghall Street followed an accelerated timetable, it's likely that Priestley would have never stumbled across his "delightful Pyrmont water"; without the brewery, it's possible that Priestley wouldn't have thrown himself into the study of gases that dominated the next decade of his research. We tend to talk about the history of ideas in terms of individual genius and broader cultural categories—the spirit of the age, the paradigm of research. But ideas happen in specific physical environments as well, environments that bring their own distinct pressures, opportunities, limitations, and happy accidents to the evolution of human understanding. Take Joseph Priestley out of Enlightenment culture, and deprive him of the scientific method, and his legendary streak no doubt disappears, or turns into something radically different. But take Priestley out of Meadow Lane, and deprive him of his hours at the brewery, and you would likely get a different story as well.

Ideas are situated in another kind of environment as well: the information network. Theoretically, it is possible to imagine good ideas happening in a vacuum—a lone Inuit scientist conjuring up breathtaking discoveries in his igloo, and then keeping them to himself. (Mendel's pea-pod experiments were not that far from this model.) But most important ideas enter the pantheon because they *circulate*. And the flow is two-way: the ideas happen in the first place because they are triggered by other people's ideas. The whole notion of intellectual circulation or flow is embedded in the word "influence" itself ("to flow into," *influere* in the original Latin). Good ideas influence, and are themselves influenced by, other ideas. They flow into each other. Different societies at different moments in history have varying patterns of circulation: compare the cloistered, stagnant information pools of the European Dark Ages to the hyperlinked, open-sourced connectivity of the Internet.

You can see in Priestley's letters to the Electricians where he and his friends fell on the circulation spectrum: every detail of every experiment relayed in the most generous, exhaustive form imaginable. The idea of proprietary secrets, of withholding information for personal gain, was unimaginable in that group. Think of the untold trillions of dollars that have been generated by the invention of soda water, and yet Priestley happily revealed his formula in letters, pamphlets, and dinner party chatter to anyone who would listen. This meant that he failed to realize the commercial potential of his invention, a decision that would have life-long repercussions for him, in that Priestley would remain, in one fashion or another, dependent on the financial patronage of other people. (A certain Johann Schweppe fared better in this regard, patenting a method of carbonating water in 1783; his namesake still enlivens gin-and-tonics to this day.) But Priestley was a compulsive sharer, and the emphasis on openness and general circulation is as consistent a theme as any in his work. The whole genesis of *The History* had been to inspire new research by conveying the current state of play in intelligible and comprehensive detail. No doubt Priestley saw farther because he stood on the shoulders of giants, but he had another crucial asset: he had a reliable postal service that let him share his ideas with giants. That reliability had its limits, however. Information networks are shaped not only by their speed and connectivity but also by their *security*. At three points in Priestley's life, crucial events would unfold precisely because a letter or batch of letters had been stolen or had somehow fallen into the wrong hands—a plot twist that recurs through the epistolary novels of the period. It's not simply the speed of information that shapes the flow of ideas in a given society, it's also how vulnerable that information is to attack or misappropriation.

Thinking about Priestley's streak in the context of information networks takes us all the way back to that fateful meeting at the London Coffee House. The open circulation of ideas was practically the founding credo of the Club of

Honest Whigs, and of eighteenth-century coffeehouse culture in general. With the university system languishing amid archaic traditions, and corporate R&D labs still on the distant horizon, the public space of the coffeehouse served as the central hub of innovation in British society. How much of the Enlightenment do we owe to coffee? Most of the epic developments in England between 1650 and 1800 that still warrant a mention in the history textbooks have a coffeehouse lurking at some crucial juncture in their story. The restoration of Charles II, Newton's theory of gravity, the South Sea Bubble—they all came about, in part, because England had developed a taste for coffee, and a fondness for the kind of informal networking and shoptalk that the coffeehouse enabled. Lloyd's of London was once just Edward Lloyd's coffeehouse, until the shipowners and merchants started clustering there, and collectively invented the modern insurance company. You can't underestimate the impact that the Club of Honest Whigs had on Priestley's subsequent streak, precisely because he was able to plug in to an existing network of relationships and collaborations that the coffeehouse environment facilitated. Not just because there were learned men of science sitting around the table—more formal institutions like the Royal Society supplied comparable gatherings—but also because the coffeehouse culture was cross-disciplinary by nature, the conversations freely roaming from electricity, to the abuses of Parliament, to the fate of dissenting churches.

The rise of coffeehouse culture influenced more than

just the information networks of the Enlightenment; it also transformed the neurochemical networks in the brains of all those newfound coffee-drinkers. Coffee is a stimulant that has been clinically proven to improve cognitive function—particularly for memory-related tasks—during the first cup or two. Increase the amount of "smart" drugs flowing through individual brains, and the collective intelligence of the culture will become smarter, if enough people get hooked. Create enough caffeine-abusers in your society and you'll be statistically more likely to launch an Age of Reason. That may itself sound like the self-justifying fantasy of a longtime coffee-drinker, but to connect coffee plausibly to the Age of Enlightenment you have to consider the context of recreational drug abuse in seventeenth-century Europe. Coffee-drinkers are not necessarily smarter, in the long run, than those who abstain from caffeine. (Even if they are smarter for that first cup.) But when coffee originally arrived as a mass phenomenon in the mid-1600s, it was not seducing a culture of perfect sobriety. It was replacing alcohol as the daytime drug of choice. The historian Tom Standage writes in his ingenious *A History of the World in Six Glasses*:

The impact of the introduction of coffee into Europe during the seventeenth century was particularly noticeable since the most common beverages of the time, even at breakfast, were weak "small beer" and wine. . . . Those who drank coffee instead of alcohol began the day alert and stimulated, rather than relaxed and mildly inebriated, and

the quality and quantity of their work improved. . . . Western Europe began to emerge from an alcoholic haze that
had lasted for centuries.

Emerging from that centuries-long bender, armed with a
belief in the scientific method and the conviction, inherited
from Newtonian physics, that simple laws could be unearthed
beneath complex behavior, the networked, caffeinated minds
of the eighteenth century found themselves in a universe
that was ripe for discovery. The everyday world was teem-
ing with mysterious phenomena—air, fire, animals, plants,
rocks, weather—that had never before been probed with
the conceptual tools of the scientific method. This sense of
terra incognita also helps explain why Priestley could be so
innovative in so many different disciplines, and why Enlightenment culture in general spawned so many distinct paradigm shifts. Amateur dabblers could make transformative
scientific discoveries because the history of each field was an
embarrassing lineage of conjecture and superstition. Every
discipline was suddenly new again. Priestley said it best in
the introduction to his *History*:

> In electricity, in particular, there is a greatest room to
> make new discoveries. It is a field but just opened, and
> requires no great stock of particular preparatory knowl-
> edge; so that any person who is tolerably well versed in
> experimental philosophy may presently be upon a level
> with the most experienced electricians.

If Priestley and his comrades unearthed an amazing trove of scientific treasure during these exceptional decades, it was at least in part because the soil was so shallow.

But to speak of soil in this context is to mix elemental metaphors. Priestley's two great discoveries from this period were made of air, not earth. One of them—by far the more celebrated of the two—revolutionized chemistry, though Priestley blundered spectacularly in interpreting his findings. But the other one he got right.

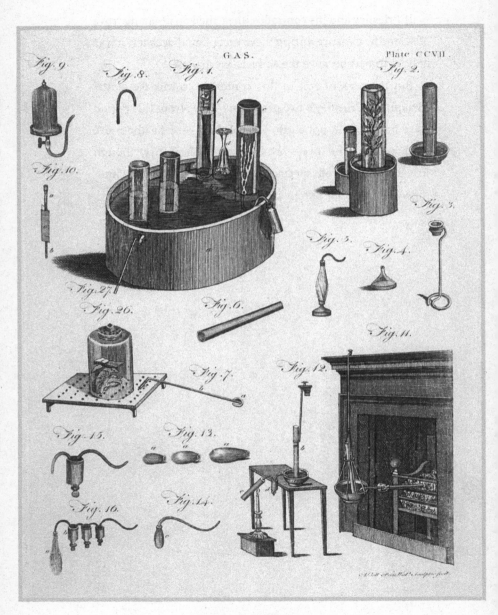

GAS. Plate CCVII.

PRIESTLEY'S TOOLS

CHAPTER TWO

# Rose and Nightshade

*August 1771*

*Leeds*

IT ARRIVED THE WAY SO MANY GOOD IDEAS do, through a brilliant mistake.

As a child growing up in rural Yorkshire, Priestley had amused himself with the slightly sadistic pastime of trapping spiders in sealed glass jars and observing how long it would take the poor creatures to perish. This hobby made such an impression on Priestley's brother Timothy that he mentioned it prominently in his funeral oration for Joseph, as evidence of his eleven-year-old brother's early aptitude for science. The fact that organisms would invariably expire given a finite supply of air was well known to little boys and scientists alike. But the mechanism behind this process was a mystery. Did the creatures somehow exhaust the air they were breathing—in which case, what was left in the jar? Or were they poisoning their environment with some invisible substance they released? Or was some other factor at work? Strangely, the air in the jar didn't visibly change after the animal's final convulsions, though it did have one distinct, and puzzling, new attribute: a lit candle would invariably flicker and die in it.

The Priestleys had finally moved into the minister's house at Bansinghall Street, where Joseph set up a home laboratory that borrowed quite a bit of its essential gear from Mary Priestley's kitchen. Even at Bansinghall Street, beer continued to play an amusing side role in Priestley's research: the mice and frogs he sacrificed in the name of science were often housed in beer glasses that he had pilfered from Mary's cabinets. The most important contraption in his laboratory was the "pneumatic trough"—a device for capturing and manipulating gases first developed by Stephen Hayles fifty years before. Priestley's first trough was Mary's laundry sink, though his accelerating research and their growing family would soon necessitate that Priestley construct his own pneumatic troughs, custom designed for his research needs. Priestley tinkered with the design of the trough constantly, but ultimately settled on a rectangular wooden box, two feet long and nine inches deep. At one end of the trough, he built a shelf, with "orifices" cut into it large enough to admit a small tube. Priestley would fill the trough with water—or, if he was working with water-soluble gases, with mercury. With the levels of water or mercury kept just above the shelf, Priestley could place his glass vessels on the shelf and conduct an amazing variety of experiments on the air they contained. The key to the trough was that the water on the bottom of the vessel at once sealed the gas, but at the same time was permeable enough to allow Priestley to insert things into the vessel. He could generate a gas in another container, and then pump it into the vessel on the shelf; or incinerate a material contained in the vessel by using a "burning lens," which concentrated the sun's rays enough to set fire to most flammable substances. For the many experiments that involved putting a live mouse into a sealed container, he grabbed the mouse by the neck, swiftly passed the creature through the water into the container, and placed them on the shelf.

Sometime in the late spring of 1771, Priestley decided to try a new twist on his childhood experiment. If animals died swiftly in a sealed jar, how long would it take a plant to suffer the same fate? It was obvious that living things couldn't survive in such an environment for long—the question was, how long? Could a plant outlast a mouse or a frog? Or would it prove more feeble in the contained environment of the jar? He went out into the garden and pulled a small mint plant from the ground. (Priestley always referred to it as a "sprig" of mint, but it appears to have been an entire plant, given his references to its stalk and roots.) He placed the mint in a glass jar that he had inverted over the pneumatic trough. And he waited, patiently, for the plant to expire.

Benjamin Franklin had paid the Priestleys a visit on May 23, within a matter of days of Priestley's decision to isolate a sprig of mint in a glass. Franklin had been traveling through northern England, enjoying a bit of industrial tourism with a few acquaintances. They had inspected the water-driven saws and polishers at the marble mills of Blakewell, floated down the Duke of Bridgewater's canal in Manchester, and descended into the cramped coal mines that lay at the canal's far end. They saw Matthew Boulton's famous Soho ironworks in Birmingham, an overwhelming glimpse of a bizarrely mechanized future: "The work of a button," Franklin's companion Jonathan Williams noted in his journal, "has 5 or 6 branches in it each of which is performed in a second of Time. He likewise works plated Goods—Watch Rings and all manner of hard ware all of which is performed by Machinery in such a Manner that Children and Women perform the greatest part of it."

In the midst of this dizzying new world of furnaces and factory floors and canals blasted through the sides of mountains, Priestley's home lab in Leeds must have seemed like an idyll. We do not know if Priestley shared his curiosity about the mint with Franklin. Williams merely notes that Priestley "made some very pretty Electrical Experiments and some on the different properties of different kinds of Air."

In the years to come, as Priestley's network of friends started to include many of the industrial magnates that Franklin had visited on his northern expedition, Priestley's lab became increasingly populated by tools that had been explicitly designed and manufactured for the needs of his research. But in his early years in Leeds there is a wonderful sense of improvisation and bricolage to Priestley's equipment, what we would now call "hacking": taking tools designed by other people for other purposes, and creatively repurposing them for your own needs.

We know so much about Priestley's gear because he compulsively shared the details of his contrivances, first with fellow Royal Society members and Honest Whigs, and then in his published works. When Priestley eventually wrapped all of his chemistry experiments into a six-volume opus, he devoted a hundred pages at the beginning to an exhaustive inventory of the tools he had used to revolutionize chemistry. Volume one began with a foldout illustration that captured the kit in loving detail, the vials, jars, and beakers lined up as if for a family portrait.

What's so striking in this image is the spirit of total openness that pervades it. There is no magician's reserve in Priestley's cabinet of wonders. He truly wants you to see everything, in mind-numbing detail, to the extent that some passages begin to sound more like self-assembly instructions for some impossibly complicated household appliance:

When I want to admit a particular kind of air to any thing that will not bear wetting, and yet cannot be conveniently put into a phial, and especially if it be in the form of a powder, and must be placed upon a stand (as in those experiments in which the focus of a burning mirror is to be thrown upon it) I first exhaust a receiver, in which it is previously placed; and having a glass tube, bended for the purpose, as in Pl. II. Fig. 14, I screw it to the item of a transfer of the air-pump on which the receiver had been exhausted.

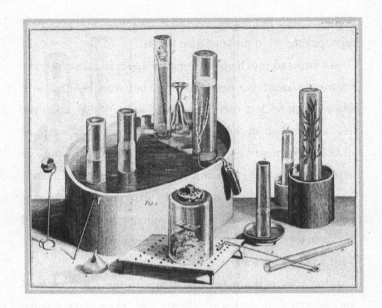

Part of this compulsive sharing no doubt comes from the fact that one of Priestley's great skills as a scientist was his inventiveness with tools. He was a hacker, not a theoretician, and so it made sense to showcase his technical innovations alongside the scientific ideas they generated. But there was a higher purpose that drove Priestley to document his techniques in such meticulous detail: the information network. Priestley's whole model of progress was built on the premise that ideas had to move, to *circulate*, for them to turn into better ideas. This is what led him to expose his technological apparatus in such detail, and what led him, on numerous occasions, to publish experimental data without fully vetting it first. It was a sensibility that he shared with Franklin, who, in a letter to Collinson in 1753, ended a long summary of his electricity experiments with the lines:

These Thoughts, my dear Friend, are many of them crude and hasty, and if I were merely ambitious of acquiring some Reputation in Philosophy, I ought to keep them by me, 'till corrected and improved by Time and farther Experience. But since even short Hints, and imperfect Experiments in any new Branch of Science, being communicated, have oftentimes a good Effect, in exciting the attention of the Ingenious to the Subject, and so becoming the Occasion of more exact disquisitions (as I before observed) and more compleat Discoveries, you are at Liberty to communicate this Paper to whom you please; it being of more Importance that Knowledge should increase, than that your Friend should be thought an accurate Philosopher.

"Exciting the attentions of the ingenious"—this was Priestley's mission in a nutshell. It defeated the whole point of the enterprise to write a book about a scientific advance, without sharing all the paths followed—and all the gear assembled—to reach that vista.

Sometimes, the false turns along those paths proved to be the most productive ones. When Priestley decided to seal up his mint in a confined jar, he fully expected it to wilt and die in a matter of days or weeks. But when he returned to the plant in June, something strange and unexpected happened. The plant had stubbornly refused to die.

The plant was not affected any otherwise than was the necessary consequence of its continued situation; for plants growing in several other kinds of air, were all affected in the very same manner. Every succession of leaves was more diminished in size than the preceding. . . . The root decayed, and the stalk also, beginning from the root; and yet the plant continued to grow upward, drawing its nourishment through a black and rotten stem.

Priestley's expectations had been entirely incorrect: in fact, the determined sprig of mint continued growing all summer long. And there were other mysteries. A candle would readily burn in the jar alongside the mint. A mouse placed inside the jar with the plant could survive happily for ten minutes, while a mouse placed in a plant-free jar in which another mouse had previously expired would begin to convulse within seconds. Somehow the plant was disabling whatever it was that snuffed out the candle and suffocated the mouse.

And here we find ourselves at the fault line of the classic Kuhnian paradigm shift, an older continent of understanding colliding with some unknown landmass. Data emerge that somehow challenge the dominant model, either by producing results that defy the expectations of the model, or by producing results that are so strange that the dominant model no longer seems relevant. It is not entirely clear from the historical record how conscious Priestley was of the full implications of what he had observed. He wrote Franklin in late summer with an account of a new discovery—but the original letter has been lost, and so we don't know with certainty that Priestley was reporting on his mint experiments. All we know is that Franklin forwarded Priestley's news along to Canton with a brief note:

> I have just received the enclos'd from Dr. Priestly. And as it contains an Account of a new Discovery of his, which is very curious, and, if it holds, will open a new Field of Knowledge, I send it to you immediately. Please to communicate it to Dr. Price when he returns.

It would have been difficult for Priestley, contemplating that tenacious sprig of mint in the lab on Bansinghall Street, to perceive that a Kuhnian revolution was at hand, not just because the concept didn't exist yet, but more important because there was no "dominant paradigm" for him to overturn. The study of air itself had only begun to blossom as a science in the past century, with Robert Boyle's work on the compression and expansion of air in the late 1600s, and Black's more recent work on carbon dioxide. Before Boyle and Black, there was little reason to think there was anything to investigate: the world was filled with stuff—people, animals, planets, sprigs of mint—and then there was the nothingness between all the stuff. Why would you study nothingness when there was such a vast supply of stuff to explain? There wasn't a problem in the nothingness that needed explaining. A cycle of negative reinforcement arose: the lack of a clear problem kept the questions at bay, and the lack of questions left the problems as invisible as the air itself. As Priestley once wrote of Newton, "[he] had very little knowledge of *air*, so he had few doubts concerning it."

So the question is: Where did the doubts come from? Why did the problem of air become visible at that specific point in time? Why were Priestley, Boyle, and Black able to see the question clearly enough to begin trying to answer it? There were 800 million human beings on the planet in 1770, every single one of them utterly dependent on air. Why Priestley, Boyle, and Black over everyone else?

One way to answer that question is through the lens of technological history. They were able to explore the problem because they had new tools. The air pumps designed by Otto von Guericke and Boyle (the latter in collaboration with his assistant, Robert Hooke, in the mid-1600s) were as essential to Priestley's lab in Leeds as the electrical machines had been to his Warrington investigations. It was almost impossible to do experiments without being able to move air around in a controlled manner, just as it was impossible to explore electricity without a reliable means of generating it.

In a way, the air pump had enabled the entire field of pneumatic chemistry in the seventeenth century by showing, indirectly, that there was something to study in the first place. If air was simply the empty space between things, what was there to investigate? But the air pump allowed you to remove all the air from a confined space, and thus create a vacuum, which behaved markedly differently from common air, even though air and absence of air were visually indistinguishable. Bells wouldn't ring in a vacuum, and candles were extinguished. Von Guericke discovered that a metal sphere composed of two parts would seal tightly shut if you evacuated the air between them. Thus the air pump not only helped justify the study of air itself, but also enabled one of the great spectacles of early Enlightenment science.

The following engraving shows the legendary demonstration of the Magdeburg Sphere, which von Guericke presented before Ferdinand III to much amazement: two eight-horse teams attempt—and, spectacularly, fail—to separate the two hemispheres that have been sealed together by the force of a vacuum.

Luftpumpe: Experiment mit Guerikes Magdeburger Halbkugeln.
Faksimile aus: Otto von Guerikes Experimenta. Amsterdam 1672.

When we think of technological advances powering scientific discovery, the image that conventionally comes to mind is a specifically visual one: tools that expand the range of our vision, that let us literally see the object of study with new clarity, or peer into new levels of the very distant, the very small. Think of the impact that the telescope had on early physics, or the microscope on bacteriology. But new ways of seeing are not always crucial to discovery. The air pump didn't allow you to see the vacuum, because of course there was nothing to see; but it did allow you to see it indirectly, in the force that held the Magdeburg Sphere together despite all that horsepower. Priestley was two centuries too early to see the molecules bouncing off one another in his beer glasses. But he had another, equally important, technological breakthrough at his disposal: he could *measure* those molecules, or at least the gas they collectively formed. He had thermometers that could register changes in temperature (plus, crucially, a standard unit for describing those changes). And he had scales for measuring changes in weight that were a thousand times more accurate than the scales da Vinci built three centuries earlier.

This is a standard pattern in the history of science: when tools for measuring increase their precision by orders of magnitude, new paradigms often emerge, because the newfound accuracy reveals anomalies that had gone undetected. One of the crucial benefits of increasing the accuracy of scales is that it suddenly became possible to measure things that had almost no weight. Black's discovery of fixed air, and its perplexing mixture with common air, would have been impossible without the state-of-the-art scales he employed in his experiments. The whole inquiry had begun when Black heated a quantity of "magnesia alba," and discovered that it lost a minuscule amount of weight in the process—a difference that would have been imperceptible using older scales. The shift in weight suggested that something was escaping from the magnesia into the air. By then running comparable experiments, heating a wide array of substances, Black was able to accurately determine the weight of carbon dioxide, and consequently prove the existence of the gas. It weighs, therefore it is.

ALL OF THIS helps us understand why the whole question of air was suddenly conceivable in Priestley's era, and why Priestley and his contemporaries were able to start solving the problem the way they did. The question of air was "in the air" not for any vague, spirit-of-the-age reasons, nor because a solitary genius had experienced a heroic epiphany. Air had become an interesting problem in large part because a handful of technologies had shed light on that most invisible of substances. The mountain was lifting the explorers higher, and in part the mountain was being moved by new tools: pumps, thermometers, scales.

But that is only a partial answer, because to explain what brought Priestley to that lab in Leeds, what compelled him to put that sprig of mint in the jar, you also have to ask the question: Why Priestley? Why not someone else? Why not Franklin or some other Honest Whig, working within the same technological regime as Priestley and Black?

This is where we normally get to the accidents of biography, the random churn of coincidences and personal anecdote, driven by both nature and nurture: *He just happened to study with an influential mentor who got him interested in the field.* Or: *He just happened to be born with some outlandish cognitive gift that let him see farther and deeper than his rivals.* Or, more comically: *He just happened to be sitting under that apple tree.*

Yet there must be recognizable streams that run beneath all that surface turbulence. In trying to answer the question of how to keep climbing the mountain, are there principles we can find on the biographical scale that can potentially help us climb other peaks? Can we learn something useful from Priestley the *individual*, from his sensibility or temperament?

Perhaps the most important factor—and the most neglected in the modern canon of how-to books on innovation—is the simple fact that Priestley was following a long hunch, one that he'd been exploring in a casual way for thirty-odd years, ever since he'd bottled up that first spider with his brother Timothy. He'd had a hunch that there was something intriguing in the whole question of why things died when you cut off their air supply, even if he didn't have the conceptual tools to solve the mystery, or even to explain why the problem seemed so intriguing in the first place. It was that hunch that led him to explore the Jakes and Nell Brewery, that brought him back to pneumatic chemistry after his immersion in electricity. Priestley's memoirs and correspondence reveal that he had ruminated on the cycle of noxious and wholesome air "for a long time" before launching into a systematic study of it, after the move to Leeds.

The idea that hunches are crucial to scientific breakthrough is nothing new, of course. What's interesting about Priestley is not that he had a hunch, but rather that he had the intelligence and the leisure time to let that hunch lurk in the background for thirty years, growing and evolving and connecting with each new milestone in Priestley's career. We know that epiphanies are a myth of popular science, that ideas don't just fall out of the sky, or leap out of our subconscious. But we don't yet recognize how slow in developing most good ideas are, how they often need to remain dormant as intuitive hunches for decades before they flower. Chance favors the prepared mind, and Priestley had been preparing for thirty years. We talk about great ideas using the language of flashes and instant revelation, but most great ideas happen on the scale of generations, not seconds. (Think of the almost glacial pace that characterized Darwin's "discovery" of natural selection.) Most great ideas grow the way Priestley's did, starting with some childhood obsession, struggling through an extended adolescence of random collisions and false starts, and finally blooming decades after they first took root.

This pattern of long cultivation holds true for Priestley's thinking across the wide spectrum of his interests. The notes on the distortions of Christianity that he filed away in his drawer in Needham in the 1750s would emerge twenty years later as a cohesive and brilliant dismantling of contemporary Christian beliefs. His hunches about restructuring the educational system first appeared in his curriculum at

Warrington, then animated the introductory chapter of *The History and Present State of Electricity*, and would ultimately play a key role in his friendship with Thomas Jefferson in the closing years of Priestley's life.

That hunches so often work this way makes intuitive sense, given the biological structure of the human brain. Ideas are built out of self-exciting networks of neurons, clusters of clusters, with each group associated with some shade of a thought or memory or emotion. When we think of a certain concept, or experience some new form of stimulus, a complex network of neuronal groups switches on in synchrony. (Priestley knew nothing about neurons, of course, but he subscribed to a generalized version of this associationist theory that he learned from the British philosopher David Hartley, whose model of cognitive "vibrations" anticipated the modern theory of neuronal association.) Priestley puts a sprig of mint under a glass in a makeshift lab in Leeds, and a hundred clusters light up in his brain: the memory of his brother Timothy and the spiders; the smell of mint in the garden at Warrington; the bad air bubbling over the vats at the brewery. Each time those associations are triggered together, the connection between them strengthens, making it more likely that they fire together as an ensemble the next time around, when some new stimulus triggers part of the network.

The shape of that idea as it forms looks nothing like the shape that intellectual history has traditionally given it. It is not the radical leap of the epiphany:

Confusion > Lightbulb > Clarity

Nor is it the oppositional ladder of the dialectic:

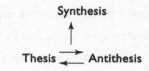

Instead, the true shape of an idea forming looks much more like this:

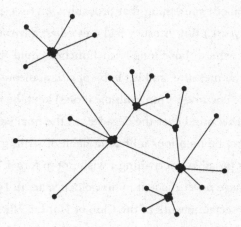

That network shape is one of the reasons why external information networks (the coffeehouse, the Internet) are so crucial to the process of innovation, because those networks so often supply new connections that the solo inventor wouldn't have stumbled across on his or her own. But the long life span of the hunch suggests another crucial dimension here: it is not just the inventor's social network that matters, but the specific way in which the inventor networks with his own past selves, his or her ability to keep old ideas and associations alive in the mind. If great ideas usually arrive in fragments, a partial cluster of neurons, then part of the secret to having great ideas lies in creating a working environment where those fragments are nurtured and sustained over time. This obviously poses some difficulty in modern work environments, with deadlines and quarterly reports and annual job reviews. (The typical middle manager doesn't respond favorably to news that an employee has a hunch about something that probably won't see results for twenty years.) But Priestley had created an environment for himself where those long-term hunches could thrive with almost no pressure, and his habit of simultaneously writing multiple documents (on multiple topics) kept the fragments alive in his mind over the decades. In the final pages of his memoirs, he mentions a lifelong habit of writing down "as soon as possible, every thing I wish not to forget." Priestley might have never made it to his golden years in Leeds without the social network of the Club of Honest Whigs. But he also had a knack for "socializing" with his own ideas.

ONE FINAL ELEMENT of Priestley's approach gave him a distinct advantage in scaling the mountain: his research style was uniquely suited for the problem he was wrestling with. At this early stage in its development, pneumatic chemistry was a field that happened to be highly receptive to Priestley's characteristic approach—take dozens of minerals or plants or organisms and subject them to an endless series of experimental variations: burning, heating, bottling up. If he had been attempting to solve the riddle of universal gravitation or natural selection, his methodology would have been useless. But the mystery of air turned out to be a problem that you could productively tackle with a laundry basin, a few beer glasses, and a gift for imagining new combinations. He had put mice and spiders into a glass jar and watched them die. So why not a plant?

But what happened next was Priestley's real stroke of combinatorial genius. After he had convinced himself that the mint in the glass was surviving despite its confinement, he decided to make a simple, but essential, modification to the experiment, one that took only a matter of seconds to engineer. On August 17, 1771, after a summer of analyzing those preternaturally healthy bottled sprigs of mint, Priestley took a thin wire and attached a "small bit of candle" to the end, and twisted the wire so that the candle end was turned upward. He lit the wick, and placed the flame next to a new sprig of mint, its roots floating in a pool of water. Then he slowly lowered a glass cylinder over the plant and the flame, and waited for the candle to burn through the supply of air in the container. Priestley knew that a mouse or spider placed into such an environment would be dead in seconds, since the candle had burned through all that was life-sustaining in "good" air. But would the plant survive? When the flame died out, he pulled the wire through the water at the base of the glass, leaving the sprig alone in the glass with no wholesome air whatsoever.

On August 27, Priestley revisited the mint in the glass. Ten days before, a flame had been snuffed out by the lack of wholesome air in the vessel, and during that period, no new air had entered the glass. Priestley knew from experience that an empty glass left in that state would be completely inhospitable to flame, as well as to any organism confined there. But when he went to light a candle in the glass, he found that "it burned perfectly well in it."

This was genuine news. His first experiment had shown that plants failed to exhaust or poison the atmosphere the way living creatures did, but that flame burning next to the sprig of mint suggested a far more radical proposition: that plants were restoring something fundamental to the air, or they were creating the air itself. He repeated the experiment "eight or ten times in the remainder of the summer . . . without the least variation in the event." The possibilities sent Priestley into a furious run of new configurations:

> Several times I divided the quantity of air in which the candle had burned out, into two parts, and putting the plant into one of them, left the other in the same exposure, contained, also, in a glass vessel immersed in water, but without any plant; and never failed to find, that a candle would burn in the former, but not in the latter. . . . I generally found that five or six days were sufficient to restore this air, when the plant was in its vigour; whereas

I have kept this kind of air in glass vessels, immersed in water many months, without being able to perceive that the least alteration had been made in it. I have also tried a great variety of experiments upon it, as by condensing, rarefying, exposing to the light and heat, &c. and throwing into it the effluvia of many different substances, but without any effect.

By the fall of 1771, Priestley was confident enough in his results to begin sharing the news with the Honest Whigs. "You may depend on the account I sent you of my experiments on the restoration of air made noxious by animals breathing it or putrefying it, which I sent to Dr. Franklin," he wrote to Price on October 3. "Air in which candles have burnt out is also restored by the same means." (He wrote Price again three weeks later to reiterate this point.) By the summer of 1772, Priestley had cycled through a series of different plants to confirm that the restorative effect was not somehow specific to mint. He began with a sprig of balm, which performed admirably. He then began to worry that the "aromatic effluvia" of those two plants was somehow the culprit, and so he tried the experiment with the "offensive"-smelling weed groundsel. Of all the plants he put under the glass, spinach proved to be the most effective at restoring the atmosphere inside the glass. In one experiment a spinach plant was able to fill the glass with combustible air in only two days.

Franklin returned for another visit to Leeds in June of 1772, this time bringing John Pringle, the Scottish physician who would soon be elected president of the Royal Society. Priestley gave them the full tour of his experiments with restoring air, and the visit seems to have energized him all over again about the importance of what he had discovered. On July 1, he wrote to Franklin:

I presume that by this time you are arrived in London, and I am willing to take the first opportunity of informing you, that I have never been so busy, or so successful in making experiments, as since I had the pleasure of seeing you at Leeds.

I have fully satisfied myself that air rendered in the highest degree noxious by breathing is restored by sprigs of mint growing in it. You will probably remember the flourishing state in which you saw one of my plants. I put a mouse [in] the air in which it was growing on the saturday after you went, which was seven days after it was put in, and it continued in it five minutes without shewing any sign of uneasiness, and was taken out quite strong and vigorous, when a mouse died after being not two seconds in a part of the same original quantity of air, which had stood in the same exposure without a plant in it. The same mouse also that lived so well in the restored air, was barely recoverable after being not more than one second in the other. I have also had another instance of a mouse living 14 minutes, without being at all hurt, in little more than two ounce measures of another quantity of noxious air in which a plant had grown.

We know with remarkable precision the sequence of experiments that Priestley conducted in this pursuit, thanks to the flow of letters and to Priestley's memoirs. We know the exact dates of many of the experiments, and the exact plants or animals he placed into his vessels. We know which ones lived and which ones died. And we can perceive through his first-person accounts his rising sense of excitement about what he had uncovered. But what is more obscure to us, looking back two centuries later, is how quickly Priestley grasped the full consequences of his experiment. To the untrained eye, it looked like nothing: a plant growing in a glass. Even to a natural philosopher, it might have seemed little more than a local curiosity: a few cubic inches of air created by a sprig of mint. But in that small parlor trick lay a whole new way of thinking about the planet itself, and its capacity for sustaining life. There was a *system* lurking in the glass that was a microcosm of a vast system that had been evolving on Earth for two billion years. Did Priestley have a hunch about this broader scale, too?

What Priestley had stumbled across is now much more than a hunch. We know that the gas that Priestley was observing was dioxygen, otherwise known as "free oxygen," or $O_2$, a molecule formed by the union of two oxygen atoms. While oxygen is the third most common element in the universe, we know that free oxygen was exceedingly rare in the Earth's initial atmosphere, until roughly two billion years ago, when an ancestor of modern cyanobacteria hit upon a photosynthetic process that used the energy from the sun to extract hydrogen from the abundant supply of water on the planet. That metabolic strategy was spectacularly successful—the organism quickly covered the surface of the planet—but it had a pollution problem: it expelled free oxygen as a waste product. During this period, now known as the Proterozoic, the oxygen content of the atmosphere exploded from 0.0001 percent to roughly 3 percent, beginning its long march to the current levels of 21 percent. (Even today, Earth's atmosphere is actually dominated by nitrogen, which makes up 78 percent of its overall volume; other gases, like argon and carbon dioxide, constitute less than a single percent.) The massive increase of oxygen in the atmosphere triggered what has been called "by far the greatest pollution crisis the earth has ever endured," destroying countless microbes for whom the cocktail of sunlight and oxygen was deadly.

In time, though, organisms evolved that thrived in an oxygen-heavy environment. We are their descendants. The invention of photosynthesis created a radically different atmosphere for Earth—an artificial bubble created by the plants, at first lethal, and then, over time, life-sustaining, as a whole new family of organisms discovered the possibilities of aerobic respiration, through the evolution of mitochondrial power plants that used oxygen to produce energy. Without those evolutionary innovations, and without the continued production of oxygen by plants and cyanobacteria, the human race would cease to exist, along with the rest of the aerobes.

Of course this immense vista—reaching back billions of years, and down to the microscopic world of bacteria and molecules—would have been almost entirely off-limits to Priestley and Franklin. But both men had a hunch that something profound was lurking in the mint's survival. The first indication of that hunch that has survived in the archives comes in a note from Franklin to Priestley after the June visit.

That the vegetable creation should restore the air which is spoiled by the animal part of it, looks like a rational system, and seems to be of a piece with the rest. Thus fire purifies water all the world over. It purifies it by distillation, when it raises it in vapours, and lets it fall in rain; and farther still by filtration, when, keeping it fluid, it suffers that rain to percolate the earth. We knew before, that putrid animal substances were converted into sweet vegetables, when mixed with the earth, and applied as manure; and now, it seems, that the same putrid substances, mixed with the air, have a similar effect. The strong thriving state of your mint in putrid air seems to shew that the air is mended by taking something from it, and not by adding to it.

In this last hypothesis, Franklin had it half right: the plant was taking and adding at the same time, producing oxygen and absorbing carbon dioxide. But his instincts about the fundamental concept were uncanny: the mint's capacity for rejuvenating "putrid" air was part of a larger system that extended far beyond an isolated laundry sink in Leeds. Franklin saw the whole story almost immediately: this discovery of

Priestley's was a key to understanding the cycle of life on Earth.

Had Priestley made that leap before, or did he need Franklin to complete the thought? The truth is we don't know, but there is a clear sense in the intonation of Franklin's note that suggests he is offering a fresh analysis of his friend's experiment, making new connections and not simply parroting something that Priestley already had told him in the Leeds laboratory. Franklin had his flaws, but obliviousness to his friends' achievements was not among them, and everything in their correspondence suggests that Franklin had come to consider Priestley his peer, if not his superior, as a scientist. Perhaps Priestley had rushed through the cabinet of wonders, and hadn't dwelt on the ramifications of his mint experiment as fully as one would expect. But given the prominence that it plays in the letters around this period, it seems entirely reasonable to assume that Priestley gave the mint experiment center stage during Franklin's visit. Priestley himself saw fit to quote directly from Franklin's musings in his initial published accounts of the experiment, which would seem to corroborate the notion that the broader, synthetic view of Priestley's discovery originated with Franklin.

If Franklin was indeed the first to propose the wider ramifications of Priestley's experiment, it would be a fitting continuation of their intellectual duet: Franklin created Priestley the scientist; Priestley popularized the legend of Franklin the daring electrician; and now Franklin was helping Priestley grasp the full significance of his discovery.

In his letter to Priestley, Franklin even managed to trace the implications of the restored air all the way up to an embryonic version of "green" politics:

> I hope this will give some check to the rage of destroying trees that grow near houses, which has accompanied our late improvements in gardening, from an opinion of their being unwholesome. I am certain, from long observation, that there is nothing unhealthy in the air of woods; for we Americans have every where our country habitations in the midst of woods, and no people on earth enjoy better health, or are more prolific.

These surviving letters between Priestley and Franklin give us front row seats to one of history's more elusive dramas: intellectual landmasses shifting underfoot thanks to a *conversation* between two people. We can see here the first stirrings of a genuinely new way of thinking about life on Earth and our role in that system. The air we breathe is not some unalienable fact of life on Earth, like gravity or magnetism, but is rather something that is specifically manufactured by plants. And that manufacture is itself part of a vast, interconnected system that links animals, plants, and invisible gases in a "rational" flow. And the choices we make as humans—destroying trees that grow near houses, for instance—can have a dangerous impact on that flow, if the core participants in the system aren't properly appreciated and protected. In discovering how Mother Nature had invented our atmosphere, Franklin and Priestley were inventing something just as profound: the ecosystems view of the world.

In the book that he would eventually publish, *Experiments and Observations on Different Kinds of Air*, Priestley spelled out the global implications in clear language:

> Once any quantity of air has been rendered noxious by animals breathing in it as long as they could, I do not know that any methods have been discovered of rendering it fit for breathing again. It is evident, however, that there must be some provision in nature for this purpose, as well as for that of rendering the air fit for sustaining flame; for without [it] the whole mass of the atmosphere would, in time, become unfit for the purpose of animal life; and yet there is no reason to think that it is, at present, at all less fit for respiration than it has ever been. I flatter myself, however, that I have hit upon one of the methods employed by nature for this great purpose. How many others there may be, I cannot tell.

By the fall, the Honest Whigs were abuzz with Priestley's discovery. Negotiations ensued within the Royal Society and by November the Society voted to award him the Copley Medal, the most prestigious scientific prize of its day, "on account of the many curious and useful Experiments contained in his observations on different kinds of Air." In receiving the prize, Priestley was joining the ranks of his friends Canton and Franklin, who had three medals between them. Only five years after they had encouraged him to turn his experimental hobbies into a serious vocation, Priestley had reached the highest pinnacle of scientific achievement. Sir John Pringle, newly elected president of the Society, gave an unusually long address in presenting the medal, explaining why Priestley's contributions were so valuable. He placed special emphasis on the mint in the glass, and the vast system of life it helped explain:

From these discoveries we are assured, that no vegetable grows in vain, but that from the oak of the forest to the grass of the field, every individual plant is serviceable to mankind; if not always distinguished by some private virtue, yet making a part of the whole which cleanses and purifies our atmosphere. In this the fragrant rose and deadly nightshade co-operate; nor is the herbage, nor the woods that flourish in the most remote and unpeopled regions unprofitable to us, nor we to them; considering how constantly the winds convey to them our vitiated air, for our relief, and for their nourishment.

Pringle twisted Franklin's rational system into a more human-centric rendition, with "every individual plant serviceable to mankind," but even with that distortion, the scope of Priestley's discovery comes through vividly in the language:

I present you with this medal, the palm and laurel of this community, as a faithful and unfading testimonial of their regard, and of the just sense they have of your merit, and of the persevering industry with which you have promoted the views, and thereby the honour, of this Society. And, in their behalf, I must earnestly request you to continue those liberal and valuable inquiries, whether by further prosecuting this subject, probably not exhausted, or by investigating the nature of some other of the subtle fluids of the universe.

THERE WOULD BE many "subtle fluids" to investigate in the coming years. Priestley would be the first to identify ten of them, including hydrogen chloride, ammonia, sulfur dioxide, and silicon fluoride. But his most celebrated—and contested—discovery would come nearly two years after the Copley Medal.

Priestley's meteoric rise to prominence as a scientist— along with his political writings—had attracted the attention of William Petty, Earl of Shelburne, former secretary of state and arguably the most intellectually nimble and inquisitive political figure in Britain. (Shelburne's Irish roots and liberal politics also made him one of the least popular.) In late 1772, the earl had proposed an arrangement whereby Priestley would maintain Shelburne's library, educate his two sons, and advise on subjects and materials currently being debated in Parliament. In turn, Shelburne would house the Priestleys and their children in far grander style than they had ever been accustomed to, spending the winter in a town house near Shelburne's residence in Berkeley Square, and the rest of the year at Bowood, the family estate in Calne, Wiltshire, in the southwest. Happy with his relative freedom and extraordinary run of success in Leeds, Priestley spent months weighing the decision. He wrote to Franklin for advice, and Franklin suggested what was then a novel approach to resolving such an issue:

My Way is, to divide half a Sheet of Paper by a Line into two Columns, writing over the one Pro, and over the other Con. Then during three or four Days Consideration I put down under the different Heads short Hints of the different Motives that at different Times occur to me for or against the Measure. When I have thus got them all together in one View, I endeavour to estimate their respective Weights; and where I find two, one on each side, that seem equal, I strike them both out: If I find a Reason pro equal to some two Reasons con, I strike out the three. If I judge some two Reasons con equal to some three Reasons pro, I strike out the five; and thus proceeding I find at length where the Ballance lies; and if after a Day or two of farther Consideration nothing new that is of Importance occurs on either side, I come to a Determination accordingly.

Ultimately, Priestley agreed to Shelburne's plan. Mary packed up their Leeds house, and the growing family—baby William had been born the year before—moved to Wiltshire in the summer of 1773. Priestley's improvised kitchen laboratory was replaced by a much more opulent setting: a laboratory in the newly constructed orangery on Shelburne's estate. The lab was next door to Bowood's imposing library, and looked out on a verdant lawn gently sloping down toward a small lake. (The grounds had been designed by the legendary landscape architect Lancelot "Capability" Brown.) Joseph and Mary had not exactly entered English high society, but for the first time in their lives, they were down the hall from it. Mary was largely unimpressed by her firsthand view of the upper classes. One story has Shelburne arriving to welcome them at their new house in Calne, and finding Mary on a ladder, industriously papering the walls. Joseph apologized for their not providing a more gracious welcome, but Mary quickly dismissed her husband's proprieties. "Lord Shelburne is a statesman," she said, "and knows that people are best employed in doing their duty." Later she would observe candidly to Shelburne, "I find the conduct of the upper so exactly like that of the lower classes that I am thankful I was born in the middle."

The quality of Priestley's tools was improved under Shelburne's patronage. Perhaps the most important was a burning glass he acquired shortly after the move, reportedly the former property of the Grand Duke Cosimo III of Tuscany. The glass was a twelve-inch convex lens, with a focal point of twenty inches that concentrated the sun's rays with great intensity and precision. Like that eleven-year-old boy trapping spiders in jars, Priestley set about with his new gadget to burn as many substances as he could possibly imagine.

What happened next may not be as famous as other eureka stories in the scientific canon, but as a case study in the complexities of intellectual history it may be the most analyzed "discovery" on record. In part this is because there is a dispute at its core, a question of precedence that is not easily resolved, and that rivalry makes for a rich narrative study. But it has also seen so much churn because it exemplifies the blurriness that so often accompanies paradigm shifts. Kuhn tells the story early on in his *Structure of Scientific Revolutions*, and published a longer version of it in the journal *Science*. Since then, it has become its own sort of experimental laboratory for theories about intellectual progress, a place where scientists could use competing models of innovation to test their hypotheses.

The facts are simple enough. In early August of 1774, Priestley turned his lens on mercury calx, the ash that forms when mercury is heated in air. It produced a gas that behaved surprisingly like nitrous air (our nitric oxide), in that candles appeared to blaze with unusual intensity in its presence. This baffled Priestley, because the mercury shouldn't have had any nitrous air in it. A friend procured purer samples and Priestley tried the experiment again. To his astonishment, the candle burned even brighter.

Shortly thereafter, Priestley left on a long European tour with Shelburne. He ruminated over the burning flame the entire trip, and at a fateful dinner in Paris—wonderfully captured in Joe Jackson's *A World on Fire*—he gave a riveting account of his experiments to an audience of *philosophes*, among them Antoine Lavoisier, who would soon be Priestley's rival, and who would eventually complete the chemical revolution that Priestley had initiated in his Leeds laboratory. It was a classic case of Priestley's inveterate openness: at a time when British and French spies actively infiltrated the industrial and scientific labs in both countries, Priestley sits down to dinner with the scientific intelligentsia of France and happily spills the beans about his most exciting new experiment. ("I never make the least secret of any thing that I observe," he would explain later, in his description of the conversation.)

When Priestley returned from the European excursion (earlier than planned, having tired of France before Shelburne did), he quickly launched into an in-depth study of this strange new air. He devised new experiments that generated even purer samples, and the more he probed the air, the more it seemed to differ from nitrous air. Slowly, Priestley began to think that he had produced common air, in which case it should support animal respiration. On March 8, 1775, he put a mouse in a glass with two ounces of air generated from the mercury calx. The mouse suffered no immediate discomfort, just as Priestley expected. But then something very odd happened. A mouse trapped in a vessel with ordinary atmospheric air would last fifteen minutes before collapsing. But the mouse that Priestley had trapped in the jar with his new air somehow survived for thirty minutes. It might have survived even longer: "Though it was taken out seemingly dead," Priestley wrote, "it appeared to have been only exceedingly chilled, for, upon being held to the fire, it presently revived and appeared not to have received any harm from the experiment."

He ran the mouse experiment multiple times in the ensuing days, reducing the amount of air available to each mouse, and each time finding the creatures strangely unfazed by a quantity of air that should have killed them in five minutes. Priestley mulled these strange facts in his head compulsively through sleepless nights, waking early each morning to try a new variation. Eventually, he mustered up the courage to employ himself as a test subject and inhale the miraculous new air himself:

> The feeling of it to my lungs was not sensibly different from that of common air; but I fancied that my breast felt peculiarly light and easy for some time afterwards. Who can tell but that, in time, this pure air may become a fashionable article in luxury. Hitherto only two mice and myself had had the privilege of breathing it.

That first breath forced Priestley, at last, to confront an astonishing truth. The ordinary atmosphere that sustained life on Earth could be improved. There was an air purer than common air. Two billion years after the cyanobacteria began pumping the Earth's atmosphere full with the stuff, Joseph Priestley had discovered $O_2$.

OF ALL PRIESTLEY'S accomplishments, all the books and ideas and experiments, all the world-changing conversations that ran through his career, the discovery of oxygen conventionally ranks at the very pinnacle of his lifework. The *Encyclopædia Britannica* entry on Priestley devotes nearly a fifth of its text to the oxygen story. Wikipedia mentions it in the second sentence on its Priestley page.

But the true narrative is more complicated than that easy declarative sentence—Priestley discovered oxygen—which is why so many scholars have dissected the particulars of this story. Priestley's breakthrough illustrates the fuzzy boundaries of scientific discovery. "Discovering" oxygen is not like "discovering" the Dead Sea Scrolls or some other unique object that has a clear identity and has been undeniably hidden for ages. It is closer to, say, discovering America: the meaning of the phrase depends entirely on the perspective and values you bring to the issue.

We know definitively from the records of their experiments that other scientists had isolated pure oxygen before Priestley took his burning glass to the mercury calx, but in each case the investigator had failed to realize the significance of what he'd done. (With one exception, to which we will turn in a moment.) Even Priestley appears to have isolated the gas in earlier experiments. What mattered was not that Priestley produced $O_2$, but that he realized that he had done something unusual, and then convincingly proved that it was a more rarified subset of common air. (Ever the practical chemist, Priestley even managed to throw in a teaser about his discovery becoming a "fashionable article" someday. He had invented soda water five years before; now he was pointing the way toward the oxygen bar.) Priestley had stumbled on the gas in one of his classic prepared accidents, but he had possessed the good sense to notice the anomaly of the flame burning brighter than expected, and he'd had the time and tenacity to explore further variations in the subsequent months.

The problem is that someone else had made comparable explorations before. The Swedish chemist Carl Scheele had isolated oxygen using a number of substances, including mercury calx, between 1771 and 1772. He called it "fire air" because of its combustible nature. He also demonstrated that common air was a mixture of two distinct gases, the fire air of oxygen, and what he dubbed the "foul air" of nitrogen. But Scheele failed to publish his findings until 1777, long after Priestley had been credited with the breakthrough.

When Priestley described his discovery, in Book IV of his *Experiments on Air*, he introduced the section with an open admission of the role of randomness in his work—even including a subtle dig at the theoretical, synthetic mode of Newton and his followers:

More is owing to what we call *chance*, that is philosophically speaking, to the observation of *events arising from unknown causes*, than to any proper *design*, or preconceived theory in this business. This does not appear in the works of those who write *synthetically* upon these subjects; but would, I doubt not, appear very strikingly in those who are the most celebrated for their philosophical acumen, did they write *analytically* and ingenuously.

It's a valid observation, given Priestley's chaotic method and his general aversion to theorizing, made even more valid by the tremendous run of success he'd just enjoyed. But there was a catch lurking in those offhand dismissals of "preconceived theories." Priestley himself was trapped in a preconceived theory, one that would prove to be almost entirely unfounded, though he clung to it for the rest of his life.

This was not a theory hiding in the shadows; Priestley seared it directly into the name he gave his pure air: *dephlogisticated air.*

That awkward name came from the closest thing to a dominant research paradigm in the nebulous field of pneumatic chemistry: the phlogiston theory, one of the all-time classics in the history of human error. Phlogiston was an attempt to explain the age-old mystery of why things burned. (The term derives from the ancient Greek word for "fire.") First proposed by the German chemist Johann Joachim Becher in the late 1600s, it was refined into a working theory by Becher's pupil, Georg Ernst Stahl, who proposed in 1716 that all substances capable of burning possessed a substance called phlogiston that was released into the air during combustion.

When the flame of a burning substance goes out, the air was considered to be "phlogisticated"—having absorbed so much of the magic ingredient of combustion that nothing remained to burn.

We now know that the phlogiston theory had things almost exactly backward, though most of the leading chemists before Priestley—including both Black and Scheele—failed to see the flaws in it and labored happily within its framework. In truth, when things burn in common air, something is being *extracted* from the air, not the reverse: oxygen molecules are bonding in the heat of combustion with whatever happens to be on fire. This is what we now call oxidation. When the air loses too many oxygen molecules to support the oxidation process, the flame goes out.

Priestley, alas, was on the wrong end of the phlogiston paradigm, and so when he happened upon an air in which flames burned more brightly than common air, he interpreted his findings using the conceptual framework of the existing paradigm. Breathable air that also exacerbated combustion was, logically, air that had been entirely emptied of phlogiston. (Or, put another way, it was air primed to be filled with phlogiston.) Within the rules of that conceptual system, Priestley's dephlogisticated air was a fitting, if ungainly, appellation. Unfortunately, the rules of that system were fundamentally flawed.

Seeing around the flaw in the model was once again made possible by technological advances in measurement. A few chemists had noted the puzzling fact that some burned substances weighed slightly more than they did before combustion, seemingly contradicting the premise that they were releasing phlogiston into the air. But like most anomalies in the decades before a paradigm shift, those uncomfortable observations were largely swept under the rug, in part because the weight gain was so minuscule.

To Antoine Lavoisier, however, that additional weight was a mystery that could not be dismissed. Inspired by Priestley's dinner-table account of his inventive experiments, but equally appalled by the Englishman's lack of theoretical rigor, Lavoisier embarked on a series of experiments that utilized his unrivaled skills with a balance. ("It can be taken as an axiom," Lavoisier wrote, "that in every operation an equal quantity of matter exists both before and after the operation.") His measurements led him outside the blinders of the phlogiston theory, and by 1776 he announced his finding that atmospheric air was one-fourth composed of "pure air . . . which Mr. Priestley has very wrongly called dephlogisticated air." The historian Joe Jackson describes it well:

Burning added weight: there was a union, shown by the most sensitive balances in Europe. There was not loss of the mysterious phlogiston. . . . All chemical changes obeyed the law of the indestructibility of matter. There were no ghosts in the process, no ether escaping notice of his scales. In the chemical change of burning, nothing was gained or lost, even in the vaporous air.

By the next year, Lavoisier was ready to give this "pure air" its scientific name. He called it oxygen.

Priestley's "discovery" of oxygen turns out to be far more vexed than the standard short-form biographies suggest. He was not the first to identify the gas, and he did not give it its enduring scientific name. The name he did affix to his discovery betrayed a fundamental misunderstanding of the basic chemistry of oxygen. No one contests the fact that he published his findings before Scheele, of course; and there is no doubt he played an essential role in leading Lavoisier to his more nuanced understanding of the gas. But the simple fact is that he was neither first, nor the most accurate, participant in the discovery of oxygen. Kuhn makes a related point in *The Structure of Scientific Revolutions* when he wonders how one can responsibly date the discovery of oxygen:

> Ignoring Scheele, we can safely say that oxygen had not been discovered before 1774, and we would probably also say that it had been discovered by 1777 or shortly thereafter. But within those limits or others like them any attempt to date the discovery must inevitably be arbitrary because discovering a new sort of phenomenon is necessarily a complex event.

What is puzzling here is not that Priestley should receive the popular acclaim for a discovery that was not entirely his; the history of exploration—whether intellectual or geographic—is ripe with false attributions and contested claims of priority. What's strange is that Priestley should be so widely recognized for his oxygen experiments of 1774–75, and yet the mint experiments of 1771–72 are often mentioned only in passing in accounts of his scientific career. (The *Britannica* entry on Priestley barely mentions the mint experiment.) Both were foundational insights that led to world-changing ideas that rippled through science and society. But there is no dispute over the mint experiments; as far as we know, he was genuinely the first to discover that breathable air was a concoction of plants, and with Franklin's help he was able to grasp and describe the far-reaching consequences that process would have on our understanding of Earth's environment. He reached that point on the mountain before anyone else, and made no missteps in his ascent. So why is he so often celebrated for a climb where he didn't actually make it to the peak?

The answer to this riddle lies in one central fact: the new science unleashed by Priestley's mint experiment took two centuries to evolve. What Priestley had hit upon was not a simple element, like oxygen, or a fundamental law, like gravity. It was, instead, a *system*, a flow of energy and molecular change. Priestley had a hand in filling out other key parts of the system as well. He connected the metabolic flow of plant respiration with the energy needs of animals in a paper published in 1776, "Observations on Respiration and the Use of the Blood." Priestley's argument was, naturally, couched in the language of phlogiston, but it was the first to suggest that there was some essential transfer of energy involved in the contact between air and blood in the lungs. In 1778, Priestley noticed that some kind of "green matter" was spontaneously forming in glasses of pump water; he noted suggestively that it required sunlight to emerge, though for a time he denied that the mysterious substance was "vegetable" in nature. The Dutch biologist Jan Ingenhousz would turn Priestley's speculations into a more rigorous proof of the energy transfer of photosynthesis, showing that a single leaf was capable of transforming sunlight into breathable air. By the end of the decade, Priestley had helped sketch out the first draft of the cycle of life on Earth: plants convert the energy of light into chemical energy, releasing oxygen into the atmosphere and absorbing carbon dioxide; animals power themselves through the energy stored in plant tissue and oxygen itself, releasing carbon dioxide as a waste product. Priestley never presented these insights as a unified system in his published work, and of course the confusion about phlogiston undermined his thinking at several key points. But in the years to come, the connective power of that system became more and more visible. In the middle of the next century, the German chemist Justus von Liebig shed important light on the nutrient cycles that drive all dynamic ecosystems. The revolutions in microbiology at the end of the nineteenth century suggested for the first time the productive role that bacteria might play in breaking down organic compounds for further reuse. The first detailed analysis of food webs—documenting the flow of energy through a population of plants, animals, and microorganisms—were sketched out by the British zoologist Charles Elton in the 1920s. Yet the word "ecosystem" wasn't even coined until the 1930s, when an Oxford botanist named Arthur Tansley asked a colleague to come up with a name for the complex interactions between organisms and their physical environments.

Like the hunch that helped bring it about, the fundamental premise behind Priestley's mint experiment took a long time to bloom into a mature science—almost two centuries, in fact. Part of that slow evolution can be attributed to the consilient structure of ecosystem science: it is a discipline that by nature is built out of the layered interactions between multiple fields of expertise, each operating on distinct scales. For the ecosystem to work as a practical object of study, you need microbiologists to explain the role of bacteria in decomposition; geologists and chemists to explain the chemical weathering of rocks; molecular biologists to explain the energy transactions of mitochondria; zoologists and botanists to identify the food webs that form between species; climatologists and atmospheric physicists to analyze the global weather patterns that shape every ecosystem on the planet. Priestley had laid the cornerstone for that amazing body of knowledge, but the building itself didn't become visible for a hundred and fifty years.

And so the legend of Priestley the scientist accumulated around his troubled discovery of oxygen, because the discovery was contemporaneous with the science it helped inaugurate. By the 1780s, the "chemical revolution" was in full swing, ignited in large part by Lavoisier's *Méthode de nomenclature chimique*, published in 1787, the founding text of modern chemistry, which established for the first time a standard nomenclature and classification system for the core elements, such as oxygen, nitrogen, mercury, and hydrogen. A new science needs its origin stories, and Priestley had undeniably been there at the beginning. Despite the Copley Medal and Franklin's enthusiasm, the mint in the glass faded into the background. In time, it would mark the origin of a new science, too, but by that point Joseph Priestley had entered the pantheon as the discoverer of oxygen, albeit one with an asterisk.

IF FRANKLIN had played a deciding role in Priestley's Leeds experiments, forces conspired to make him but a distant observer of his friend's discovery of dephlogisticated air. In late 1772, a member of Parliament secretly passed Franklin a packet of letters written by the Massachusetts governor, Thomas Hutchinson. The letters talked openly of restricting "English liberties" in the colonies in order to suppress the growing rebellion. Franklin sent them back to his friend Thomas Cushing in Boston—apparently thinking that somehow they would help ease tensions with England. Instead, the letters were published to much outrage among the colonists. As the American protests grew in intensity, there was fierce speculation in London over who had sent the Hutchinson letters back to Boston. Franklin ultimately revealed that he had been the culprit, and in early 1774, he was denounced in front of the Privy Council, in the famous Cockpit chamber in Whitehall. The solicitor-general Alexander Wedderburn claimed Franklin's duplicity had "forfeited all the respect of societies and men," his mind "possessed by the idea of a Great American Republic." There were catcalls from the audience against Franklin, but he had allies in the room as well: Priestley was in attendance, watching his fellow Honest Whig suffer perhaps the greatest public humiliation of his life. The Cockpit was so crowded that Priestley was forced to stand the entire session, next to their mutual friend Edmund Burke. (Ironically, when the political furies turned against Priestley fifteen years later, Burke would be one of Priestley's key antagonists.) Along with Lord Shelburne and a young Jeremy Bentham, they formed a small band of Franklin supporters in an otherwise hostile crowd. They were, to a man, appalled by the ferocity of Wedderburn's attack. Shelburne later called it "scurrilous invective." After the session ended, Wedderburn approached Priestley in the antechamber to extend his greetings; Priestley turned his back on the solicitor-general and immediately marched out to the street in protest.

The next morning, Franklin had breakfast with Priestley and insisted that he had no regrets about his actions. Hours later, a note arrived, informing Franklin that he had been stripped of his cherished position as postmaster general for his role in inciting the colonial uprising—as well as for circulating purloined letters, something of a faux pas for the postmaster general. Fearing arrest or the wrath of an angry mob of patriots, Franklin took a boat downriver to a friend's house in Chelsea, where he kept out of the public eye for a few weeks. He threatened to leave England for good, but ended up staying for another year, despite an almost complete end to his dealings with the ministry. During that summer, which Priestley spent in Wiltshire with Lord Shelburne, the surviving correspondence between Franklin and Priestley drops down to one or two letters, and there is no record of Priestley's sharing his early investigations into pure air with his longtime collaborator.

The winter of 1774–75 gave Priestley and Franklin one last opportunity to revive their close relationship. The Priestleys moved to London for the winter, and he and Franklin dined together nearly every night, often in the company of the Honest Whigs. The shoptalk had turned almost exclusively to politics by that point, given the disintegrating bond between England and her transatlantic colonies. It is not stretching matters to suggest that Priestley's thinking suffered from the political distractions of his old friend. Two years before, after witnessing the mint growing in the jar, Franklin had suggested in a letter that "the air is mended by taking something from it, and not by adding to it." *Taking*, not *adding*. If only Franklin had suggested a parallel hunch to Priestley in 1774, and steered him clear of phlogiston's magnetic pull, the chemical revolution might have played out quite differently.

By March of 1775, Franklin realized that the time had

finally come to pledge his allegiance to his native land, and he booked passage on a packet ship leaving Portsmouth on March 11, bound for Philadelphia. He spent the entirety of his final day in London—the last he would spend there as a subject of the British crown—with his friend Joseph Priestley. They read through a package of American newspapers that had just arrived, surveying the colonial responses to the Boston Port Act, which had established a blockade against the Boston harbor as a reply to the Tea Party. "As he read the addresses to the inhabitants of Boston," Priestley later recalled, ". . . the tears trickled down his cheeks."

A year later, when Priestley published the first edition of his *Observations and Experiments on Different Kinds of Air*, the defining account of his golden years as an experimental scientist, he would invoke the loss of his great friend in the opening pages.

The greatest success in [politics] seldom extends farther than one particular country, and one particular age; whereas a successful pursuit of science makes a man the benefactor of all mankind, and every age.

Then he quoted a private letter from the Italian philosopher Francesco Beccaria:

*Mi spiace che il mondo politico, ch' è pur tanto passeggero, rubbi il grande Franklin al mondo della natura, che non sa ne cambiare, ne mancare.*

[I am sorry that the political world, which is so very transitory, should take the great Franklin from the world of nature, which can never change, or fail.]


Priestley couldn't have known it then, transcribing those words in the safety of his lab in Calne or Berkeley Square, but in time, the world of politics would take him as well.

The Carboniferous Era

## Chapter Three

# Intermezzo:
# An Island of Coal

*300 Million B.C.*

*Pangaea*

In 1877, THE FRENCH PALEONTOLOGIST Charles Brongniart began excavating fossils from the coal measures near the town of Commentry, in central France. It was a promising spot for a fossil hunt; originally a lake several miles long, the site had been bordered by a marshland where streams from the surrounding hills drained down, depositing plants and insect life in the swamp. Brongniart was only eighteen at the time he discovered the site, and he would work the Commentry quarry for almost twenty years, nearly the entire span of his adult life. (He died at the age of forty.) During that period he unearthed a spectacular array of fossils, most of them dating back 300 million years, older than the first dinosaurs.

Many of the fossils that Brongniart uncovered shared a defining characteristic: compared with their modern equivalents, they were massive. He discovered ferns the size of oak trees, and flies as big as birds. In 1880 he unearthed his most startling find: a monster dragonfly with a wingspan of 63 centimeters. Brongniart named it *Meganeura* in the paper he published about his discovery in 1894. The original fossil can be seen today in the Museum of Natural History in Paris. Subsequent fossils have been discovered with a wingspan of more than 75 centimeters.

*Meganeura* was not alone. Paleontologists worldwide soon discovered that giantism was a prevailing trend between 350 and 300 million B.C., a period now called the Carboniferous era. Like some strange Brobdingnagian natural history exhibit, the landscape of the Carboniferous was populated by foot-long spiders and millipedes, and water scorpions the size of a small boy. The plant life was even more spectacular. Club mosses growing in damp forests towered above the swampland below, reaching heights of 130 feet, five hundred times taller than their modern descendants. Horsetails and rushes that now top out at around four feet regularly reached the height of a five-story building. Early conifers sprouted leaves that were three feet long.

The planetary fad for giantism didn't last. The dinosaurs evolved immense body plans in the coming ages, of course, but by 250 million B.C. the rest of the biosphere had largely retreated back to the scale that we now see on Earth. But that pattern was distinct enough that it presented a tantalizing mystery: just as the Cambrian explosion raised the question of why life suddenly grew so diverse, the Carboniferous age raised the question of why life suddenly grew so *big*, and how it managed to survive with such exaggerated proportions. *Meganeura* shouldn't have been able to fly, given its size. The respiratory systems of modern insects and reptiles wouldn't be able to generate enough energy to support a body plan that was ten times their current size. And yet somehow the giants of the Carboniferous managed to thrive in that exaggerated state for a hundred million years.

ALMOST EXACTLY two centuries after Priestley first explained the mystery of breathable air, scientists began to piece together the puzzle of *Meganeura*, and when they did, the process that Joseph Priestley had first observed in his Leeds laboratory turned out to be central to the story. The giants of the Carboniferous illuminate the enduring power of Priestley's original mint experiment, the long flame of associations and insights that came out of that original spark.

Priestley and Franklin's hunch that plant life was central to the planet's production of breathable air first approached scientific consensus in the late 1960s, after two physicists, Lloyd Berkner and Lauriston Marshall, proposed in a seminal paper that the vast majority of atmospheric oxygen originated in photosynthesis. The "natural" level of oxygen on Earth was less than 1 percent; the 20.7 percent levels we enjoy as respiring mammals was an artificial state, engineered by the evolutionary breakthrough that began with cyanobacteria billions of years ago. The scarcity of oxygen before the evolution of plant life suggested one logical follow-up question: Why had oxygen levels stabilized at around 20 percent for so many millions of years? It is easy to imagine that number fluctuating more dramatically over evolutionary time: were it to drop to 10 percent, most of aerobic life would suffocate; were it to double, the combustion reactions of oxygen would engulf the planet in a worldwide inferno. So what mechanism allowed the atmosphere to regulate itself with such precision, like some kind of emergent global thermostat keeping the planet in its oxygen comfort zone?

That knack for self-regulation—also known as homeostasis—was the driving question that led James Lovelock and Lynn Margulis in the early 1970s to formulate their famous and endlessly debated Gaia Hypothesis, in which the two argued that "early after life began, it acquired control of the planetary environment and that this homeostasis by and for the biosphere has persisted ever since." That control system sought an "optimal physical and chemical environment for life on this planet."

Lovelock and Margulis began the first significant paper they published on Gaia with the story of oxygen's miraculous stability. They gave the paper the provocative title "Atmospheric Homeostasis By and For the Biosphere." *By* and *for*: these were fighting prepositions. Not only had the planet achieved some kind of sustained atmospheric balancing act, with oxygen levels maintained at optimal levels for its present biosphere, but that biosphere had somehow collectively been responsible for it, acting in its own self-interest. We accept the premise that organisms have comparable purposes in the systems that collectively keep them at homeostatic norms: our bodies stay marvelously calibrated at 98.6 degrees for a reason, and that reason is that our particular mode of staying alive is optimized for that temperature. That is one of the defining characteristics of what it means to be an organism: a system of cells and organs that are explicitly devoted to ensuring the survival of the larger group to which they belong. Each works, in the language of the original Gaia paper, as "a contrivance specifically constituted for a set of purposes." The cells that help pump blood through our bodies go to elaborate lengths to keep blood-pressure levels at an equilibrium, because stable blood pressure is important to the survival of the organism. Lovelock and Margulis saw the same principle at work on a planetary scale: the Earth itself could be seen as a single organism, with the collective behavior of every member of the biosphere contributing to its survival. It was a variation on Sir John Pringle's "no vegetable grows in vain" homily, with mankind replaced by Mother Earth. The biosphere regulates $O_2$ levels, and it does it for a reason: because stable $O_2$ levels are good for the biosphere.

A thousand holes have been punched in the Gaia Hypothesis in the three decades since Lovelock and Margulis first proposed it, and it remains an open question whether the strong claim—the Earth is an organism—has empirical merit, or even utility as a metaphor. (Lovelock and Margulis have backtracked from some of their more provocative assertions, while at the same time defending their central premise.) The weaker claim, that there are planetary systems that settle around stable states far from their "natural" equilibrium, and that life has a knack for evolving solutions that thrive in those conditions, is largely unchallenged. (It is the founding principle behind the Earth Systems perspective that we saw in the Bretherton diagram.) But wherever one falls on the spectrum of responses to Gaia, there is no contesting that it was one of the most electric and influential ideas of the late twentieth century.

One of the intriguing side effects of Gaia is that it helped trigger a multidisciplinary search to determine if oxygen levels had indeed been consistently locked in at 21 percent over the ages. In 1989, the geologists Robert Berner and Donald Canfield published a paper that described a "rock abundance" approach to measuring changing levels of oxygen in the atmosphere. By measuring the levels of carbon and sulfur in sedimentary rocks for each geological period—drawn largely from the extensive data compiled by oil companies seeking new deposits of fuel—Berner and Canfield were able to build a portrait of atmospheric oxygen dating back 600 million years. In general, Berner and Canfield's model reinforced the Gaia story: oxygen levels had been relatively stable for the last 200 million years. But the most startling finding came before that long equilibrium. The data showed a dramatic spike in oxygen levels, reaching as high as 35 percent around 300 million B.C., followed by a plunge to the borderline asphyxia of 15 percent in the Triassic era, 100 million years later. The oxygen pulse overlapped exactly with *Meganeura* and the other giants of the Carboniferous.

Since Berner and Canfield's original study appeared, dozens of papers have explored the connection between increased oxygen content and giantism, and the growing consensus is that higher oxygen concentration would support larger body plans in reptiles and insects. And the increase in atmospheric pressure that accompanies 35 percent oxygen levels would even alter the aerodynamics enough to allow *Meganeura* to take flight.

Where did all that oxygen come from? From the plants, of course. First, the plants invented the photosynthetic engine that created an oxygen-rich atmosphere billions of years ago. But at some point near the end of the Devonian age, the plants evolved the ability to generate a sturdy molecule called lignin that gave them newfound structural support, allowing them to grow to sizes never seen before on Earth. Larger plants alone might have led to an oxygen increase, but lignin may have also had a more indirect role in the spike. One popular but unproven theory argues that lignin confounded the microbial recyclers responsible for the decomposition of organic matter. Plants absorb carbon dioxide and produce oxygen through photosynthesis; decomposition plays that tape backward, as bacteria and other animals use up oxygen in breaking down the plant debris, releasing carbon dioxide in the process. Lignin may have disrupted that cycle, because the recyclers had not yet evolved the capacity to break down the molecule, creating what the paleoclimatologist David Beerling calls an episode of "global indigestion." With the decomposers handicapped by lignin's novelty, immense stockpiles of undecomposed biomass filled the swamplands and the forest floor, and the oxygen levels climbed even higher. Oxygen would not return to the 21 percent plateau until the microbes cracked the lignin code, millions of years later.

But the debris accumulated during the age of *Meganeura* did not disappear from the geological record. It simply went underground. When it ultimately resurfaced, it would transform human history every bit as dramatically as it transformed natural history the first time around.

THIS IS WHERE it is important to pause for a second and contemplate the story in its full scope. Two billion years ago, the cyanobacteria concoct a metabolic strategy that envelops the planet with oxygen. By 300 million B.C., the strategy has proven to be so successful that the Earth is literally overwhelmed with vegetation, and oxygen levels reach unprecedented heights, before stabilizing again. A long parade of events follows: dinosaurs go extinct, mammals rise, continents separate, *Homo sapiens* evolves, language appears, agriculture blooms. And then Joseph Priestley sits in a room in Leeds and watches a plant grow in a glass, and grasps—for the first time in recorded history, as far as we know—the original breakthrough that made aerobic life possible in the first place.

That sounds like a story of genius and epiphany—*a billion years of evolution and then one guy figures it all out!*—but we know that framing the story that way misses the complexity of what actually happens when great ideas come to light. We know that the mountain grows in complicated, layered ways, that Priestley was positioned to see the problem of air for many reasons: his brain and biography, his method, the technology of the day, the information networks, the scientific paradigm. But emphasizing those interacting forces—the ecosystems view of cultural achievement—doesn't take anything away from the essential magic of linking these two breakthroughs across the immense span of evolutionary time: the original invention of air, and our human understanding of the process that made that invention possible.

But there is another subterranean link that connects Priestley and the ancient cyanobacteria. All that debris that piled up during the explosion of oxygen 300 million years ago was literally lying beneath his feet as he performed his Leeds experiments, in the form the Yorkshire coal measures, part of the extensive Carboniferous layer that runs throughout northern England. The coal measures are a geological anomaly, one of the most extensive stockpiles of Carboniferous rocks ever discovered on the planet. Most of the coal measures lie in shallow beds just below the surface, though in some places they break out into open air. (Carboniferous limestone outcroppings border the peaks of the Pennine mountain range.) The disproportionate amount of nonbiodegraded organic matter trapped in the Carboniferous layer makes it an unparalleled source of fuel. Even today, 90 percent of the world's supply of coal dates back to the Carboniferous.

Britain turned out to be blessed with two happy accidents

of geology: it had an unusually large stockpile of Carboniferous fuel, and the stockpile's shallow location made it unusually accessible. (Hence the old saying about Britain being an
"island of coal.") Those Carboniferous rocks are central to
the story of why industrialization happened in England first
(and northern England, more precisely). Yes, Britain had a
technical and entrepreneurial culture that helped it exploit
all that stored energy, but without the coal measures themselves, it's entirely likely that the Industrial Revolution would
have originated somewhere else.

This is a recurring theme of human history: major
advances in civilization are almost invariably triggered by
dramatic increases in the flow of energy through society. The
birth of agriculture enabled humans to stockpile energy in
the form of domesticated plants and livestock, thus enabling
the larger population centers that evolved into the first cities. Empires became possible thanks to innovations that captured the energy required to move armies and government
officials across large distances, via the muscular energy of
horses or the harnessed wind power of ships. Industrialization took the stored energy of Carboniferous rocks and
combined it with ingenious new technology that exploited
that energy in countless ways. The result of that new energy
influx was a nation utterly transformed in little more than a
century: a tremendous increase in wealth and innovation, a
radical restructuring of the relationship between town and
country, and a whole new way of life—industrial labor—with
all the terror and trauma that entailed.

Seeing human history as a series of intensifying energy flows is one way around the classic opposition between the Great Men and Collectivist visions of history. You can tell the history of the world through the lives of individuals, or groups of individuals, and part of that explanation is no doubt true. But you can also tell that story with the humans in a supporting role, not the lead. You can tell it as the story of flows of energy: growing, subsiding, being captured, being released. Think of those flows as a vast, surging ocean, and the individual human lives of history crowded on a sailboat in that turbulent water. The humans can still steer their vessel, and exploit the waves and wind that happen to be pushing in the direction they wish to go. But the humans are largely subservient to the conditions set by those oceanic forces. If the pioneering industrialists in England hadn't hit upon the strategy of using coal to power mechanized labor, there is little doubt that some other culture would have stumbled across the same idea in the next century. But if the Carboniferous age had played out differently—no oxygen spike, no giants, no planetary indigestion—the history of modern human civilization would be radically transformed, because the dominant source of energy that powered the first wave of industrialization around the world wouldn't exist.

THERE IS A MORE speculative question here that connects us back to Priestley via another angle: To what extent do order-of-magnitude changes in energy flows affect the creation of new ideas? The anecdotal evidence would seem to suggest at least a correlation between the two: in the long sweep of history, intellectually and technologically dynamic societies tend to burn more fuel than their contemporaries. But the causal link between the flow of energy and the flow of ideas may also be more indirect. Thus far, radical increases in energy have led, almost without exception, to two long-term trends: an overall increase in wealth, and an increase in social stratification. (Most people improve their standards of living eventually, but the elites benefit disproportionately.) Those two factors growing in sync invariably produce at least one subsidiary lifestyle trend: more leisure time. And in Priestley's age at least, leisure time was where ideas happened. You can't dabble in scientific experiments when you've got to use all your cognitive resources just to put food on the table, or when you don't even have a table to put the food on. Priestley was a professional minister and educator, in that he was paid directly for those labors, but in some fundamental sense he was an amateur scientist, particularly through the first two decades of his life. Like most of his Enlightenment-era peers, he was a hobbyist where science was concerned.

The price of leisure time is ultimately paid in the currency of energy. Imagine Joseph Priestley transported back to the Dark Ages, as a village priest with a contrarian edge. Even if you could somehow magically implant in his brain all his personal knowledge of chemistry circa 1771, it's unlikely that he would have ever run the mint experiment, for the simple reason that the mint experiment took months to explore and tweak and contemplate, and only a tiny fraction of the population had that kind of spare time in those energy-poor centuries. You don't have time for hobbies when you're living hand to mouth. (Had he been a monk or a prince, things might have been different, because the monks and the princes had leisure time.) We tend to think of money encouraging innovation because it functions as an incentive, and indeed one of the legacies of the coal-powered economic revolution of the eighteenth century is that it created a scientific-industrial marketplace where good ideas could be rewarded with immense fortunes. But accumulated wealth played almost the opposite role in most Enlightenment-era science: it allowed people like Joseph Priestley to pursue scientific breakthroughs *without* the promise of financial reward. And the lack of a monetary incentive made it easier for Priestley and the Honest Whigs to share their ideas as freely as they did.

So when we attempt to answer the question of why scientific revolutions happen—why Joseph Priestley should have hit upon the secret of where breathable air comes from, and in doing so unleash a new way of thinking about the system of life on the planet—the long-zoom perspective necessarily widens beyond the immediate details of biography, past the cultural trends and technological developments, all the way out to the *longue durée* of the carbon cycle. This should be true of almost all important historical events, because energy flows are such a crucial factor in the development of human societies. But there is a beautiful symmetry in imagining Priestley's intellectual labor in this light, because he was discovering the very process that, 300 million years before, had set in motion a chain of events that ultimately afforded him the leisure time to make the discovery in the first place. The mountain of scientific understanding grew higher in part because it was sitting on a island of coal.

There is a fearful symmetry lurking in this vista, too. In the following two decades, Priestley's life would grow even more intertwined with the ancient biomass trapped in those Carboniferous-era coal deposits. That unleashed energy would propel him into the second great intellectual collaboration of his career. It would also nearly take his life.

THE LUNAR MEN

## CHAPTER FOUR

# The Wild Gas

*July 1791*
*Birmingham*

FOUR WEEKS AFTER FRANKLIN SPENT HIS
emotional final day in London with Priestley, British soldiers set off from Boston to arrest John Hancock and Samuel Adams, triggering the famous ride of Paul Revere, the
"shot heard around the world," and the astonishing retreat
of the redcoats. Franklin was still in the mid-Atlantic at that
point, but when he finally set foot in Philadelphia, he quickly
penned a letter to Priestley with his take on the news:

> You will have heard before this reaches you, of a march
> stolen by the regulars into the country by night, and of
> their *expedition* back again. They retreated 20 miles in
> [6] hours.
>
> The Governor had called the Assembly to propose Lord
> North's pacific plan; but before the time of their meeting,

began cutting of throats; You know it was said he carried the sword in one hand, and the olive branch in the other; and it seems he chose to give them a taste of the sword first. . . .

All America is exasperated by his conduct, and more firmly united than ever. The breach between the two countries is grown wider, and in danger of becoming irreparable.

I had a passage of six weeks; the weather constantly so moderate that a London wherry might have accompanied us all the way. I got home in the evening, and the next morning was unanimously chosen by the Assembly a delegate to the Congress, now sitting.

The "transitory" world of politics had once again trumped the "timeless" world of science. Franklin only had room for a brief but provocative allusion at the end of his letter. "In coming over I made a valuable philosophical discovery," he wrote, "which I shall communicate to you, when I can get a little time. At present am extremely hurried."

That valuable philosophical discovery was most likely the "gulph stream." We know Franklin had taken his water temperature measurements during that 1775 voyage, and a few days after his initial note to Priestley from Philadelphia, he began writing a new letter, recounting his involvement in the packet-ship mystery as Postmaster General. But the letter was never completed (the draft is in the Library of Congress now), and the next few missives he sent Priestley dealt almost exclusively with the state of the war and Franklin's immersion in revolutionary politics.

In early 1776, Priestley wrote Franklin:

> I lament this unhappy war, as on more serious accounts, so not a little that it renders my correspondence with you so precarious. I have had three letters from you, and have written as often; but the last, by Mr. Temple, I have been informed he could not take. What is become of it I cannot tell.

He then launched back into the world of science, mentioning his recently published *Observations on Air*, which he was including with the letter, and describing his latest round of experiments, lately focused on the circulation of blood. In the final paragraphs, he wrote:

> In one of your letters you mention your having made a valuable discovery on your passage to America, and promise to write me a particular account of it. If you ever did this, the letter has miscarried, for which I shall be sorry and the more so as I now almost despair of hearing from you any more till these troubles be settled.

There is no evidence that Franklin ever managed to relay his "valuable discovery," despite Priestley's reminders. It was a pattern that would play out through the rest of their correspondence: Franklin obsessed with the volatile state of the Revolution, unable to turn his mind back to the timeless pursuits of natural philosophy; Priestley offering support for the American cause, but then trying to shift the conversation back to the laboratory. "Though you are so much engaged in affairs of more consequence, I know it will give you some pleasure to be informed that I have been exceedingly successful in the prosecution of my experiments since the publication of my last volume," Priestley began a typical letter from late 1779. The two countries that Franklin had considered home were at war with each other, and the side he was supporting was losing. Who had time for letters about the "gulph stream" in such a context? Franklin, the world's most celebrated scientist-statesman, had, at the end of his life, become merely a statesman.

Franklin clearly grieved the loss of his natural philosophy, and in an extraordinary letter written from Passy, outside Paris, in 1782, he wrapped all that intellectual regret into a withering, near-misanthropic attack on his fellow men. Franklin begins with a homily to the importance of leisure time in scientific discovery: "I should rejoice much if I could once more recover the Leisure to search with you into the Works of Nature." But then, before the sentence can barely come to an end, he switches into a dyspeptic political mode, dividing the world again into the timeless and the transitory: "I mean the inanimate, not the animate or moral Part of them. The more I discover'd of the former, the more I admir'd them; the more I know of the latter, the more I am disgusted with them." And then Franklin launches into a sentence whose sprawling assault on humanity is matched only by its equally sprawling syntax:

> Men I find to be a Sort of Beings very badly constructed, as they are generally more easily provok'd than reconcil'd, more dispos'd to do Mischief to each other than to make Reparation, much more easily deceiv'd than undeceiv'd, and having more Pride & even Pleasure in killing than in begetting one another, for without a Blush they assemble in great Armies at Noon Day to destroy, and when they have kill'd as many as they can, they exaggerate the Number to augment the fancied Glory; but they creep into Corners or cover themselves with the Darkness of Night, when they mean to beget, as being asham'd of a virtuous Action.

Franklin the satirist appears next: "A virtuous Action it would be, and a vicious one the killing of them, if the Species were really worth producing or preserving; but of this I begin to doubt." Mindful of his much more magnanimous reader, he artfully draws the argument back to Priestley's ministry and his experiments, with a dark twist of the knife at the end, borrowed from Swift:

> I know you have no such Doubts, because in your Zeal for their Welfare, you are taking a great deal of Pains to save their Souls. Perhaps as you grow older you may look upon this as a hopeless Project, or an idle Amusement, repent

of having murdered in mephitic Air so many honest harmless Mice, and wish that to prevent Mischief you had used Boys and Girls instead of them.

Reading the letter now, it is easy to be impressed by the pyrotechnics of Franklin's style, and by his bleak view toward his own species. But the most moving words come near the end of the letter, as he settles his pen, following the outburst:

> But to be serious, my dear old Friend, I love you as much as ever, and I love all the honest Souls that meet at the London Coffeehouse. I only wonder how it happen'd that they and my other Friends in England, came to be such good Creatures in the midst of so perverse a Generation. I long to see them and you once more, and I labour for Peace with more earnestness, that I may again be happy in your sweet Society.

There is lacerating honesty in Franklin's attack on the "very badly constructed" species of man, but what is perhaps most striking in this letter is the emotional honesty in these fond words for the Honest Whigs. When you look at Franklin through the lens of his friendship with Priestley and the coffeehouse society—if you take those closing lines literally—one overwhelming thought confronts you: Franklin was, at heart, a Londoner. He made his name in Philadelphia several times over, and was revered and seduced by Paris, but there is every reason to believe that had King George been a little less aggressive with his tax policy, Franklin would have spent the last forty years of his life in London. He "labours for peace with more earnestness" so that he can return, at long last, to the sweet society of the London Coffee House. No doubt he is exaggerating his adopted-homesickness for the benefit of his old friend, but you have to willfully misread the passage to avoid the conclusion that one of Franklin's key grievances against King George was that he forced him to board that ship for Philadelphia in 1775, and leave behind the "honest souls" in the shadow of St. Paul's.

FRANKLIN MAY HAVE BEEN at the front lines, but Priestley's ideas were bound up in the American Revolution as well, in his dual capacity as scientist and political theorist. In his musings on his discovery of dephlogisticated air, he had written of its potential military uses, speculating that the new chemical techniques that he had developed might well be employed to improve the explosive power of gunpowder, or make its manufacture more efficient. He shared many of these ideas with a former Portuguese monk he had befriended named John Hyacinth Magellan. Magellan, a descendant of the famous navigator, turned out to be a spy for the French, who sent back to Paris extensive missives on the nascent scientific-industrial complex in Britain, with Priestley's research playing a starring role. Priestley himself may have shared some of his speculations with Lavoisier during their famous dinner in Paris in late 1774. As it happened, Lavoisier had just been appointed head of Louis XVI's state-subsidized Régie Royale des Poudres et Saltpêtres (the Royal Gunpowder and Saltpeter Administration), dedicated exclusively to increasing France's supply of powder. The cutting-edge ideas about combustion that Priestley and Magellan had put in his head—along with his estimable skills as a chemist—made him a brilliant choice to revitalize France's gunpowder production. By 1777, Lavoisier had increased the annual production of saltpeter to 2 million pounds, and significantly increased its explosive yield. By the early 1780s, France's saltpeter was widely considered to be the highest-quality powder in the world, propelling canonballs 50 percent farther than the British powder did.

All that stored energy—created this time by the human-driven processes of industrial chemistry, and not the carbon cycle—would eventually flow across the Atlantic, to the aid of the struggling Continental army. One of the primary reasons Franklin was so preoccupied with matters of war was the simple fact that his country didn't have enough energy on its side. At the beginning of the conflict, all thirteen colonies had between them only 80,000 pounds of gunpowder, a supply that wouldn't have lasted half a year of fighting. "Oh, that we had plenty of powder; I would then hope to see something done here for the honour of America," Nathaniel Greene wrote as he contemplated the British stronghold of Boston from his camp at Prospect Hill, north of the city, in the summer of 1775. By December of that year, Washington announced: "Our want of powder is inconceivable. A daily waste and no supply administers a gloomy prospect." Supporters of the revolutionary cause in the colonies were given a crash course in the production of gunpowder, but their concoctions were generally of poor quality, and in any event the amount of powder generated was paltry compared to the immense needs of the army.

What ultimately turned the tide were two interrelated developments, the first predicated on Priestley and Lavoisier's chemical revolution, and the second on Ben Franklin's skills as a diplomat. Lavoisier's innovations in gunpowder production gave the French a stockpile of top-quality powder. During his secret mission to France in late 1776–77, Franklin helped negotiate a pact that brought more than 200 tons of high-grade French gunpowder to the muskets of the Continental army. By 1779, more than 800 tons had been imported. That tremendous influx of stored energy changed the balance of power between the struggling colonial army and the redcoats. "By Yorktown," Joe Jackson writes, "British soldiers complained that they could not get close enough to shoot colonials before they themselves were blasted from their garters." From his laboratory in Paris, Lavoisier mused on the role of his saltpeter in the American Revolution: "It can truthfully be said that it is to those supplies that North America owes its freedom." It was typical of Lavoisier's self-important style to attribute the American victory to his own saltpeter, and no doubt he exaggerated matters in phrasing it that way. But it is impossible to imagine that freedom being won on 40 tons of mediocre powder.

Priestley had a hand in the ideas behind the colonial

struggle as well. He had published his initial work of political theory, *An Essay on the First Principles of Government,* in 1768, which expanded on Lockean liberal ideals and made some early and influential gestures toward the concept of separating church and state, though it was not nearly as widely read as his scientific publications. At Franklin's urging, Priestley published in 1774 an address to "Protestant Dissenters of All Denominations" that specifically focused on the "American Affairs." The pamphlet took a strong stand against "forg[ing] chains for America," and advised its readers to vote against all members standing for election who supported the existing policies toward the colonies "to the imminent hazard of our most valuable commerce, and of that national strength, security, and felicity which depend upon UNION and on LIBERTY." By the time Franklin set sail for Philadelphia, Priestley and Richard Price had become the most well-known British supporters of the American cause, in part provoking Samuel Johnson's famous attack on the colonial uprising, *Taxation No Tyranny.* Johnson was said to have remarked, "Ah, Priestley. An evil man, Sir. His work unsettles everything." (He was right about the second bit.) If there was a fifth column rising in support of the Americans among the intellectuals of Georgian England, its epicenter was at the London Coffee House. When Priestley wrote back after Franklin's first emotional tribute to the Honest Whigs in 1776, he thanked his old friend for his "kind remembrance" of the club, and sent a message of political solidarity: "Our zeal in the good cause is not abated."

Despite that zeal, Priestley's political worldview was still in a germinal state during the 1770s. His output of political pamphlets came to an abrupt halt in 1774, and would not start up again until the next decades. (He would ultimately write much more about the French Revolution than about the American.) This unusual reticence may have been partly attributable to sensitivities of having Lord Shelburne as a patron; while Shelburne had left the Cabinet largely because of his opposition to the king's taxation policies, he was still very much attached to the Court and Parliament, with some ambition of returning to some official office in the future. For Shelburne to have been seen as funding a vocal supporter of the colonies would have raised eyebrows. And so Priestley kept most of his explicit support for the uprising in America to the word-of-mouth networks emanating from the London Coffee House.

Priestley was less successful at subduing his religious views. In 1774, he had assisted his friend the Reverend Theophilus Lindsey in founding the first official Unitarian denomination, which openly denied the divinity of Jesus Christ and the existence of the Trinity. He published several materialist philosophical tracts during his tenure with Shelburne that questioned the notion of the soul, most markedly *Disquisitions Related to Matter and Spirit*, in 1777. These were, of course, political acts as much as they were theological ones, since the Test and Corporation Acts prohibited any dissenters from the thirty-nine articles of the Church of England from holding political office. (It was legally considered an act of high treason for a British native to say mass.) When George III lifted some of these restrictions against Catholics, while retaining them for other dissenters, Priestley appealed for a Royal audience to make the case for expanding the scope of reform, working through William Eden and Lord North. The king sent North a curt response that records his antipathy toward Priestley: "If Doctor Priestley applies to my librarian, he will have permission to see the library as other men of science have had: but I cannot think the Doctor's character as a politician or divine deserves my appearing at all."

THERE ARE TWO basic ways to look at Priestley's years at Leeds and Calne in the 1770s: either taking a contemporaneous approach, viewing his achievements in the context of their time; or, alternatively, taking the hindsight view, from the perspective afforded by our knowledge of all the events that were to come. The first view is simpler, despite all the intricacies of the tale: it's the story of a great scientist hitting his stride. The hindsight view presents a different picture: the multiple trails of Priestley's intellectual life converging for the first time, dominated by the science, to be sure, but increasingly integrated with religious and political values. In the next decade, the three paths would combine to form a mighty highway, one that would ultimately drive Priestley all the way to the New World.

We cannot fully understand Priestley—or the wider context of social change during that period, particularly among his compatriots across the pond—without appreciating the convergence of these three intellectual paths. Scientific innovation tends to be imagined as something that exists outside the public sphere of politics, or the sacred space of faith. (Recall that Kuhn barely mentions either in *The Structure of Scientific Revolutions*.) But for Priestley, these three domains were not separate compartments, but rather a kind of continuum, with new developments in each domain reinforcing and intensifying the others. When Lindsey opened his Unitarian Church, Priestley defended the move against critics who claimed it would undermine the existing religious authorities by invoking the very same principles that governed his scientific research: expose as many ideas as possible to as many minds as possible, and the system will ultimately gravitate toward truth and consensus. "[The] only method of attaining to a truly valuable agreement," he wrote, "is to promote the most perfect freedom of thinking and acting . . . in order that every point of difference may have an opportunity of being fully canvassed, not doubting but that . . . Truth will prevail, and that then a rational, firm, and truly valuable union will take place."

The critique of the soul launched during the Calne years deliberately followed the same approach that Priestley had taken in his experiments with air in the Leeds laboratory. Just as Priestley had demystified what had conventionally been called the "spirit" of mephitic air, or fermenting liquids, so would he demystify the "spirit" of human existence. These were not metaphors, strictly speaking, but elements of a connected system: the materialism that helped him isolate pure air could just as readily be applied to the theological question of the soul. The presence of a higher power in all of this wasn't somehow miraculously hovering over the human body; it lay instead in that steady widening of understanding that materialist science made possible. The progressive movement of Enlightenment science stood as the great embodiment of God's work on Earth—to Priestley a much more sensible embodiment of the divine than a man crucified almost two thousand years before. And that movement had too much force not to wipe away the political and theocratic relics that had been carried over from earlier, less sophisticated ages.

Priestley's introduction to *Observations on Air*, penned in 1774, made the connections explicit: "This rapid process of knowledge," he wrote, ". . . will, I doubt not, be the means, under God, of extirpating *all* error and prejudice, and of putting an end to all undue and usurped authority in the business of *religion*, as well as of science." To our modern ears, this is a perfectly acceptable premise, though it was daring in its day: that the ascendancy of scientific thinking would challenge the explanatory models of religion. But Priestley then goes on to make an even bolder suggestion, linking the march of scientific progress to political change, and making it clear that his own native country would not be immune: "The English hierarchy (if there be anything unsound in its constitution) has equal reason to tremble at an air pump, or an electrical machine."

Part of what makes the hindsight view so intriguing is

that we have no figure in the current intellectual landscape, in the United States at least, who fits Priestley's mold in any convincing way. Here was a man at the very front lines of scientific achievement who was simultaneously a practicing minister and theologian—and who was, by the end of the 1770s, well on his way to becoming one of the most politically charged figures of his time. He was an empiricist driven by a deep and abiding belief in God, who was simultaneously a revolutionary of the first order. In today's culture, a Venn diagram of science, politics, and faith would show no overlap, particularly if we're talking about individuals who hold radical views in all three disciplines. There are plenty of politicians with strongly religious beliefs, and plenty of clergymen who have active and influential political careers. But the vast majority of that group is conservative in its values, in the most general sense of the word: they are attempting to conserve and protect some kind of traditional order. And the scientist-politicians—Al Gore notwithstanding—are as rare as the scientist-priests.

If an echo of the intellectual chord Priestley managed to strike—political thinker, believer, scientist, radical—still somehow resonates in the American context, it is because most of those notes were played, in a slightly different configuration, by the great polymath intellectual of the revolutionary generation, Thomas Jefferson. That the two men were in such harmony was no accident, as Jefferson himself borrowed some of his dominant notes from Priestley's score. But that is getting ahead of our story.

By 1779, Priestley's controversial views had created an uncomfortable tension between himself and Lord Shelburne, who would return to public office several years later, serving as prime minister for nine months, starting in 1782. (He would negotiate the close of the Revolutionary War with Benjamin Franklin during that tenure.) When his patron suggested that he had plans to relocate Priestley to Ireland, Priestley took the proposal as indication that he had worn out his welcome. The exact details of the break remain something of a mystery. It may have been a case of Shelburne anticipating his return to public office and recognizing the potential dangers of having a radical Unitarian on his payroll as his son's tutor. Shelburne was said to blame the whole affair on Priestley's ill health. (He suffered a debilitating attack of gallstones during this period.) It may have been some kind of strange failure of communication between the two men, as the whole separation played out without any direct dismissal from Shelburne. But the most plausible interpretation comes from Priestley's biographer Robert Schofield: Shelburne had married his second wife, Louisa Fitzpatrick, in July of 1779. The daughter of the Earl of Upper Ossory, she quickly established herself as an eminent political hostess, and apparently found the Priestleys distasteful, as much for their middle-class sensibility as for Joseph's radical views. Tellingly, Shelburne attempted a rapprochement with Priestley some years later, and while the dates are not exact, it seems probable that the olive branch arrived after the death of Lady Shelburne in 1789.

The cause continues to be a matter of debate, but the effect is a matter of fact: after extensive consultation with friends, Joseph and Mary Priestley packed up the vials and air pumps and electrical machines (along with their three children) and moved to Birmingham. Priestley was giving up Berkeley Square and Calne for the heart of coal country.

Shelburne had left Priestley with an annual allowance of £150 to continue his work in Birmingham, but his financial situation was greatly compromised by the break with Shelburne. For some time, Priestley imagined that he would have to return to private tutoring to cover his family's expenses. But the threat to Priestley's valuable leisure time was quickly defused. This time around, it would not be a landed aristocrat who saw the value in supporting Priestley's work—it would be an extended group of wealthy individuals, almost all of whom had made their fortunes in the nascent Industrial Revolution. For the rest of his tenure in England, Priestley would live indirectly off the stored energy of the Carboniferous era.

The first great break for the Priestleys came when Mary's brother, the successful ironmaster John Wilkinson, secured a comfortable home for the family at Fair Hill, on the outskirts of Birmingham. The house had four main bedrooms and servants' quarters, and ample grounds for the children to explore. The upstairs floor had a long, narrow room that Priestley used as a library, though it doubled as a kind of eighteenth-century media room: Priestley would entertain children with magic lantern shows there, and harmless shocks from his electrical machine. The only flaw with Fair Hill was that it didn't include a suitable space that could be converted into a laboratory, but Priestley quickly turned that liability into a strength by constructing a separate building for his experiments, custom-tailored to his idiosyncratic needs. This was a fitting beginning to his sojourn in the Midlands, since his new coterie of industrial magnates was soon to provide him with a host of new tools designed according to his exact specifications. The Priestleys moved to Fair Hill in September of 1780, and by the end of November, the new laboratory had been completed. Priestley wrote the potter Josiah Wedgwood to report that he was ready "to do more business in a philosophical way than ever."

Still there was the matter of that £150 annuity. The Priestleys needed at least twice that to make ends meet. Even before the move to Fair Hill, Priestley had begun supplementing Shelburne's annuity by building a collection of "subscribers" who supported his work with annual contributions. The eighteenth-century concept of subscribing is one without an exact modern equivalent, falling somewhere between a magazine subscription and a charitable donation to a museum or park or university. The donation came with perks—Priestley's subscribers were sent first editions of all his writing—but the money contributed generally exceeded by a wide margin the market value of the publications. It was nice to be first in line to read Priestley's latest, of course, but one subscribed because Priestley himself was a cause worth supporting. For Priestley, subscription was a way of diversifying the patronage system; rather than tying his fortune to the whims of a single aristocrat, Priestley was assembling a broader support network to keep his ideas alive.

Unfortunately, the first round of subscriptions procured after the break with Shelburne was minimal, and one of his key supporters, the Quaker physician John Fothergill, died shortly after Priestley set out on his own. But word of Priestley's situation soon began to circulate among the Midlands industrialists, originating most likely with Wilkinson, and by early spring of 1781, a group had formed that would collectively keep Priestley in business for the next thirteen years. These were the shining lights of industrial and intellectual England outside the metropolis of London: Wilkinson; Wedgwood; the "toymaker" Matthew Boulton, whose small metal goods had become the signature export of Birmingham; James Watt, the steam-engine pioneer; and the physician, poet, and naturalist Erasmus Darwin, Charles's grandfather. The men constituted the core members of the legendary Lunar Society, Birmingham's version of the Club of Honest Whigs. The Lunaticks—as they playfully referred to themselves—had first assembled in the mid-1760s, scheduling meetings during the full moon to assist their passage home after a long night of boozy debate. The historian Jenny Uglow describes a typical session in her epic account of the society, *The Lunar Men*:

They tried to dine at two o'clock, and usually planned to stay until at least eight. The wine flowed . . . and the tables were heavy with fish and capons, Cheddar and Stilton, pies and syllabubs. At dinner the wives sometimes joined the men and the children dashed in and out. But when the meal was cleared away, out came the instruments, the plans and the models, the minerals and machines. In the house and in the workshops they talked long into the evenings.

Priestley (and Franklin) had been long-distance associates of the society since its inception, but Priestley's move to Birmingham would quickly establish him as one of the core Lunaticks. They moved their regular sessions from Sunday to Monday to reduce conflicts with the sermons Priestley had begun delivering at the New Meeting house in Birmingham. Priestley's credentials as a scientist—with Lavoisier, he was now considered the most accomplished chemist on the planet—made him a particularly valuable addition to the group, which was heavy with industrial innovation but weaker when it came to pure natural philosophy. By the 1780s, most of the society could easily afford to endow their new comrade with all the leisure time he needed. In March of 1781, Wedgwood wrote Boulton: "Our good friend, Dr. Darwin, agrees with us in the sentiment, that it would be a pity that Dr. Priestley should have any cares or cramps to interrupt him in the fine vein of experiments he is in the midst of, and is willing to devote his time to the pursuit of, for the public good." Boulton quickly wrote back with his support for the idea, suggesting that Wedgwood "manage the affair so that we may contribute our mites to so laudable a plan without the Doctor knowing anything of the matter." The discretion was not merely a question of modesty; Priestley was already controversial enough for his religious and political views that some of his supporters couldn't risk a public association with him.

Before long, Priestley was able to write to his old friend, Thomas Percival: "I am as rich as I wish to be. My sons will have employments, which I prefer to estates, under their uncles; so that I really think my lot the happiest in the world, as I can devote my whole time to useful and pleasing pursuits; and if one fails, I can fly to another."

Priestley soon fell into a routine at Fair Hill that would lead to the happiest years of his life. He rose early and worked for five or six hours on his experiments or writing projects before the midday dinner, "leaving the afternoon for visiting and company." Joseph and Mary lovingly tended to their garden on the Fair Hill grounds. It was a trifle compared to Capability Brown's immense landscape at Bowood, of course, but this was *their* garden, unlike Shelburne's lavish spread. They had four children now: their daughter Sally was seventeen when they moved to Birmingham, and her younger brother Joseph Jr. eleven; the rambunctious William was almost ten, and little Harry was just a toddler. Practically everyone who left a report of visiting Fair Hill remarked on the playful spirit of the home environment that the Priestleys had built for their family, with its impromptu magic lantern shows and tussles on the lawn. According to Joseph Jr.'s account, a day rarely went by without Priestley spending at least two hours playing games: chess, backgammon, whist. Nearly every night Joseph and Mary would play two or three matches of chess, though as the children grew older, boisterous family card games became more frequent at Fair Hill.

The one conspicious absence in this arcadia was Benjamin Franklin, who was stationed across the Channel at Passy, serving as the American ambassador to France. Despite that relative proximity, Priestley had not seen his old collaborator since Franklin's emotional last day in London in 1775. Franklin often contemplated making the journey, but his gallstones prevented it. In the summer of 1784, he wrote to Richard Price:

> I had indeed Thoughts of visting England once more, and of enjoying the great Pleasure of seeing again my Friends there: But my Malady, otherwise tolerable, is, I find, irritated by Motion in a Carriage, and I fear the Consequences of such a Journey; yet I am not quite resolv'd against it. I often think of the agreeable Evenings I used to pass with that excellent Collection of good Men, the Club at the London, and wish to be again among them. Perhaps I may pop in, some Thursday Evening when they least expect me.

Even in France, though, Franklin went out of his way to augment Priestley's reputation as a natural philosopher, lobbying to have him included (alongside Franklin himself) as one of the foreign members of the Academy of Science. When the Academy members finally agreed to it, Franklin wrote with obviously delight to Price: "I had mention'd him upon every Vacancy that has happen'd since my Residence here, and the Place has never been bestow'd more worthily."

Franklin's last voyage home to Philadelphia in 1785, which he courageously undertook at the age of seventy-nine, made it apparent to the two men that they were not likely to see each other again in person. Priestley continued sending his scientific volumes, and in one of the last surviving letters between them, Franklin offered his thanks, along with a fitting tribute to the unique experimental skills of his fellow Honest Whig. "I know of no Philosopher who starts so much good Game for the Hunters after Knowledge as you do," he wrote. "Go on and prosper."

In 1788, two years before his death, Franklin wrote to Priestley's former student Benjamin Vaughan, who had helped Franklin in negotiating the terms at the end of the war. "Remember me affectionately to good Dr. Price and the the honest heretic Dr. Priestley," he wrote. "I do not call him honest by way of distinction; for I think all the heretics I have known have been virtuous men. They have the virtue and fortitude or they would not venture to own their heresy; and they cannot afford to be deficient in any of the other virtues, as that would give advantage to their many enemies; and they have not like orthodox sinners, such a number of friends to excuse or justify them. Do not, however mistake me. It is not to my good friend's heresy that I impute his honesty. On the contrary, 'tis his honesty that has brought upon him the character of heretic."

As his long attachment to the oldest member of the revolutionary generation came to an end, Priestley forged the first links of a new American connection. During his Birmingham years, he kept to his habit of making regular pilgrimages to London, absorbing the eclectic shoptalk of the coffeehouse scene, and meeting with his old friends from the Honest Whigs. A new informal society had sprouted up around the Piccadilly store of the bookseller and publisher James Stockdale, where various London intellectuals would gather to discuss world events and theological musings. On April 19, 1786, John Adams—living in London as America's ambassador to the Court of St. James's—recorded in his diary that he had walked to "the booksellers," where he "met Dr. Priestly for the first time." They discussed the biblical descriptions of the conquest of Canaan, and the revolutionary battles in South Carolina. "I spent the Day upon the whole agreeably enough," Adams wrote, adding prophetically: "Seeds were sown, this Day, which will grow."

THE LUNAR SOCIETY contributed more than just sterling to Priestley's "useful pursuits": Wedgwood built earthenware vessels for the new laboratory; Boulton and his brothers-in-law designed metal goods, and Priestley had state-of-the-art glass instruments manufactured by a local impresario who

had built a prosperous business selling equipment for making Priestley's soda water. The arrangement was not a matter of pure charity, however. Priestley regularly advised his
Lunar Society friends on their commercial projects, examin-
ing samples of clay and iron ore for Wedgwood and Boulton,
and sharing his latest research on combustion with Watt. He
became a kind of floating, one-person R&D lab, his time and
intellect shared by multiple corporations. Priestley's chronic
intellectual openness occasionally worried his more proprietary colleagues, particularly James Watt. (Industrial secrecy
ran quite against the grain of Priestley's general propensity
to share everything with anyone who would listen.) But for
the most part, Priestley's engagement with the Lunar Society
proved a brilliant partnership, every bit as collaborative and
encouraging as the Club of Honest Whigs had been fifteen
years before.

One of the things that makes Priestley's career so interesting to us now is that his work lay at the intersection point of
four institutional models of idea production, two of which
were just emerging into a recognizable shape during his
lifetime, and two that were just beginning a long slide into
relative obscurity. Today, we take it for granted that advances
in science or technology are cultivated in two primary environments: private businesses, or public organizations like
universities or research hospitals, the latter often supported
by government funding. Priestley's move to Birmingham
planted him squarely at the origin point of the first category,
in the fusion of science, technology, and capitalism that the

new class of entrepreneur industrialists like Watt and Boulton helped bring about. Priestley had seen the second model at work firsthand during his visit to France with Shelburne in 1774; the planned, state-supported model of innovation that buttressed the work of his great chemical rival, Lavoisier.

But Priestley was only a tourist in those two soon-to-be-dominant environments; his career mostly flourished in different soil. First, there was the model of the solo, free-agent investigator—working alone in his lab, supported by a single patron or small group of patrons who refrained from meddling with his research objectives. And there was the loose connectivity of the small society—the Honest Whigs and the Lunaticks—a group of intellectual allies with different fields of expertise, sharing insight and inspiration (along with the porter and Stilton), supporting one another emotionally and, at times, financially. That Priestley would spend so much of his career happily ensconced in these less structured environments should come as no surprise to us. The amateur and the small society were the two prevailing frameworks for Enlightenment science, and they were uniquely suited for a maverick, cross-disciplinary thinker like Priestley. In the two centuries that have passed, both models have become as rare and antiquated as one of Priestley's electrical machines, replaced by the giant turbines of big industry and big government.

Priestley actually envisioned another model that "might be favourable to the increase of philosophical knowledge." He had sketched it out in the opening pages of *The History and Present State of Electricity*, at the very beginning of his scientific career, in 1767. He began by drawing attention to the existence of institutional systems like Lavoisier's Académie Française, "large incorporate societies, with funds for promoting philosophical knowledge in general." Priestley liked the idea of funding innovation, but he objected to the centralized nature of those societies. So he proposed to break them up into smaller and more nimble clusters. "Let philosophers now begin to subdivide themselves, and enter into smaller combinations," he wrote.

> Let the several companies make small funds, and appoint a director of experiments. Let every member have a right to appoint the trial of experiments in some proportion to the sum he subscribes, and let a periodical account be published of the result of them all, successful or unsuccessful. In this manner, the powers of all the members would be united and increased. Nothing would be left untried, which could be compassed at a moderate expence, and it being *one person's business* to attend to these experiments, they would be made, and reported without loss of time.

This vision is classic Priestley in the way it mirrors his own eclectic, improvisational research style. The diversity of groups, and the diversity of proposed experiments, ensures that a broad mix of interesting problems will be explored, but the accountability of the single "director of experiments" in each cluster wards off the inertia of bureaucracy that drags down so many large collaborations. Suffice it to say that this framework for innovation did not catch on the way the corporate or university model did. Whether that historical non-event is regrettable—or whether Priestley's vision was ultimately untenable in practice—is a topic for another book. What is important for our current purposes is that Priestley did come close to creating the environment he had outlined in early 1767 with his participation in the Lunar Society. There in the British Midlands, in the lab at Fair Hill, a "director of experiments" (Priestley) was supported financially by a small cluster of members, each of whom proposed different experiments to be carried out, with the results shared with, and analyzed by, the entire group. It had taken him fifteen years, but by the time Priestley settled down to work in his new lab at Fair Hill, in the fine company of the Lunar Men, he had finally established the work environment he had dreamed of as a young man.

The irony of those years at Fair Hill is that they did not turn out to be the pinnacle of Priestley's natural philosophy, despite his boast to Wedgwood. He would devise some ingenious experiments in Birmingham, and publish over a dozen papers on a typically diverse array of topics: nitrous oxide, the composition of water, and many iron- and steam-related inquiries inspired by his new environs (and the requests of his patrons and friends in the Lunar Society). He also added two new volumes to his opus *Experiments and Observations on Different Kinds of Air*. But none of his discoveries from that period

Steven
Johnson
.
162

would compare with the Leeds innovations in their ultimate social and scientific impact. There are no mint sprigs in the Fair Hill canon, and no soda water. A quiet acknowledgment of this deficiency is visible today in the center of Birmingham, in the statue of Priestley that stands in front of the Birmingham Library. The statue depicts Priestley employing his burning lens to extract pure air from mercury calx—an experiment that took place in Wiltshire, not Birmingham. Priestley spent many of his Birmingham days fighting a losing battle against Lavoisier and his critique of the phlogiston theory, drawing the Lunar Men into the debate for a time, before most of them eventually parted ways—on that one point at least—with their chief scientist. Priestley would cling to the phlogiston model for the rest of his life, despite the steady accumulation of data and expert opinion that suggested it was fatally flawed. His refusal to abandon the theory has been the subject of intense commentary over the years, and in certain accounts of the history of science, he has never been fully forgiven for the error. Even the Lunaticks wondered why their gifted colleague seemed so intent on sticking with what was clearly a losing bet.

The technological and experimental context that had served him so well during his initial investigation into the mystery of air turned out to be poorly suited to fending off the attacks on phlogiston. Priestley was a qualitative scientist, not a quantitative one. He had access to miraculously precise scales that shaped so much of Lavoisier's new chemistry, but rarely invoked measurements that exacting in his research. All the classic twists in Priestley's natural philosophy are almost existential in nature: the plant lived; the mouse died; the flame went out. Lavoisier's new chemistry, on the other hand, was the story of minutiae: this gas weighed a fraction of a gram less than this gas. Discovering that there was an air purer than pure air required the qualitative analytic skills—and improvisational style—that Priestley possessed in abundance. But defining the chemical composition of that air took a different toolkit, both mental and technological. In Kuhn's language, Priestley's skills were optimally suited for uncovering anomalies in the existing paradigm (to the extent that there was anything stable enough to be called a scientific paradigm). In this capacity, he was essential to the revolutionary science of the new chemistry, in that it was his inspired—and somewhat chaotic—explorations that first stumbled on the holes in the model, producing new facts the model couldn't explain. But when it came time to actually build new explanations, to establish the rules of play for the new paradigm, Priestley's approach ultimately undermined his efforts.

If the small group patronage model did not work quite as well in practice as it did in Priestley's theory, given his less-than-stellar record of natural philosophy during his Fair Hill years, it was not that the intellectual ecosystem in Birmingham failed to support great scientific research. You couldn't blame Priestley's phlogiston problems on his patrons in the Lunar Society, after all. If the Birmingham scene had a deleterious effect on Priestley's science, it was an indirect one, in that the new environment liberated Priestley's political and theological radicalism, which drew him into a series of distracting new controversies. Priestley could well have done his finest natural philosophy in his new lab at Fair Hill, but, like Franklin before him, the "transitory" world of politics finally pulled him into its orbit.

If you listen to Priestley's private voice, in the letters to Franklin and Price and Canton that began in the late sixties, it's clear that this radicalism was there all along. He simply didn't have an independent economic platform stable enough to support a full-throated rendition of his beliefs, though the works he did publish still managed to offend their fair share of authorities, from the king on down. But the Lunar Men gave him cover, because they were each, in their different ways, as radical as Priestley, despite the fact that some of them were captains of industry. Not just because the Lunaticks were unusually tolerant to maverick ideas, but also because their economic and political interests aligned them with Priestley's radicalism.

Here once again we find the Carboniferous age altering the course of eighteenth-century British politics. The fact that the coal measures were centered in northern England shifted the nation's economic balance of power away from the prosperous rural estates of Sussex, Essex, and Kent. The agrarian capitalism that thrived in those regions was itself a story of energy flows: the relatively balmy south of England was optimized for capturing energy directly from the sun, and so farming communities settled there in the Middle Ages, ultimately creating a thriving agrarian economy, particularly after the enclosures and improvements of the sixteenth and seventeenth centuries greatly increased the yield of the farm system. That temperate maritime climate owed its existence to the massive energy transfer of Franklin's "gulph stream," which keeps England far warmer than it should be given its distance from the equator. London, after all, lies in the same latitude as Newfoundland, where it regularly snows as late as May, and the average temperature in July peaks at 55 degrees. Without the Gulf Stream, England's green and pleasant land would be covered by snowpack six months of the year.

Those patterns of energy flow replicated themselves in the patterns of human settlement, with the population centers clustering in the sunnier and more fertile southern regions, starting in the Middle Ages. Naturally, political power also settled around these energy-rich environments; by Priestley's time, 70 percent of the House of Commons represented boroughs south of the imaginary line between Bristol and London.

The south possessed a natural environment that was better suited for extracting energy from live plants; the north for extracting energy from plants that had died 300 million years ago. But that Carboniferous energy was useless without the technology to pull it out of the ground and put it to work. Humans could live prosperously in the south with the older techniques of mass farming, but the stored energy of the north required the technologies of industrialization for it to be valuable.

When those technologies arrived, in the late seventeenth and early eighteenth centuries, the social transformation they unleashed was swift and violent. Its lucrative metal trades fueled by nearby coal deposits in South Staffordshire and Warwickshire, Birmingham would double in size every thirty years through the 1700s, creating a new class of untitled magnates like Boulton and the Wilkinsons, and a newly urban laboring class toiling on the factory floors. The demographic changes were as dramatic as any England had experienced in her history as a nation, and yet the composition of Parliament remained constant, like a fossil in a swirling sandstorm. Birmingham, the fourth-largest city in the country, with nearly 70,000 residents, did not have a single representative in Parliament. This was the great politico-economic disconnect of eighteenth-century England: the map of Parliament was based on a map of England's energy supply circa A.D. 1300. The nation was surging through the Industrial Revolution, but its political system was still trapped in its agrarian past.

For the emerging magnates of the Lunar Society, then, life in Birmingham was a sort of smaller-scale rendition of the colonies' taxation without representation. They were creating immense wealth and technological supremacy without a single Parliamentary seat. That economic and geographic situation instilled a deep-seated opposition to the archaic structures of the British establishment. Most of the Lunar Men were religious Dissenters as well, and thus doubly ostracized by the Parliamentary system. Recall Priestley's line about the

"English hierarchy" with its potentially "unsound constitution." If they had reason to "tremble at an air pump," they had even more to fear from a steam engine.

Herein lies the unique value proposition the Lunar Men saw in Joseph Priestley: as a scientist, he could improve the efficiency of their steam engines and ironworks; and as a famously prolific political *engagé*, he could fight for the reform that those booming factories had made necessary. Birmingham lay at a rare historical nexus: rapidly accumulating wealth that was simultaneously dedicated to overthrowing the status quo. No wonder, then, that Priestley's published voice grew bolder during his Birmingham years. He was riding the crest of a great dialectical wave: a massive swell of new capital headed toward the shore, intent on destroying all the ancient structures in its path.

Priestley would spend eleven years at Fair Hill, almost exactly the same duration he spent at Leeds and Calne, and in a sense, the two periods run parallel to each other. For Priestley had a second "streak" during his tenure in Birmingham in the 1780s publishing some of that turbulent decade's most influential and incendiary tracts of political and theological writing, shaping events and minds in England, America, and France. Here again the long-zoom approach turns out to be essential to understanding how this second streak came into being: Priestley's own private intellectual commons, where ideas from different disciplines were allowed to mingle and procreate; the information networks of the Lunar Society, with its comparable diversity of expertise and interest; the economics of small-group patronage, which was itself made possible by the capital accumulations of early industrialization; the Parliamentary conflicts that were ultimately shaped by energy deposits that had originally been captured from the sun before the age of dinosaurs. So when we try to answer the question of what drove Priestley up the mountain of radicalism during his sojourn in Fair Hill, we can answer the question in part by using the traditional methods of explanation: we can trace his intellectual lineage, and perform close readings of his published work and correspondence; we can describe the political pressures and conflicts of the time and explain how Priestley engaged with them. But the long view is just as essential; the wealth that the Carboniferous made possible literally paid his bills during that period; and the clash between the two energy systems—the coal deposits and the Gulf Stream—created a political climate that cultivated and nourished his radical views in crucial ways.

There is one key distinction between Priestley's two streaks, however. Priestley's hot hand in the 1770s ended with the whimper of his falling out with Shelburne. But his Fair Hill years ended with an inferno.

THE FINAL CRISIS that sent Priestley into exile was prefaced by three main controversies during the 1780s, like a series of ominous tremors leading up to a devastating earthquake. The first was the publication, in 1782, of his *History of the Corruptions of Christianity*. Originally envisioned as a supplement to the fourth edition of *Institutes of Natural and Revealed Religion*, Priestley's catalogue of all the supernatural garnish that had been layered over the original edifice of Christianity grew so extensive that he ended up publishing it as a stand-alone two-volume work. *The Corruptions* was a kind of historical deconstruction of the modern Church, isolating every instance of magic and mysticism—starting, of course, with the divinity of Jesus Christ, and the existence of a Holy Ghost—and tracing each back to the distortions of Greek and Latin theologians starting in the fourth and fifth centuries A.D., around the time of the Council of Nicea. *The Corruptions* opens with a meticulous assault on the Trinity, which takes up the first quarter of the book, then widens into a long litany of smaller abuses, the false mysticisms of the Eucharist, predestination, the immateriality of the soul, the Last Supper. The chapter on saints and angels strikes a typical note of disdain for contemporary beliefs, explaining not only the errors of the modern view, but the evolutionary path that led to those errors:

> The idolatry of the Christian church began with the deification and proper worship of Jesus Christ, but it was far from ending with it. For, from similar causes, Christians were soon led to pay an undue respect to men of eminent worth and sanctity, which at length terminated in as proper a worship of them, as that which the heathens had paid to their heroes and demigods, addressing prayer to them, in

the same manner as to the Supreme Being himself. The same undue veneration led them also to a superstitious respect for their *relics*, the places where they had lived, their pictures and images.

Priestley had no patience for the millions of Christians—in England and elsewhere—who deified the saints. They were indistinguishable from "those who bowed down to wood and stone, in the times of Paganism."

Despite those repudiations, Priestley took great care in *The Corruptions* to present the book as a *defense* of the Christian faith, restoring it to the original values of Jesus himself, and the "primitive" fathers who worshiped a single god and had no room for supernatural explanations of life on Earth. "If I have succeeded in this investigation," he explained in the preface, "this *historical method* will be found to be one of the most satisfactory modes of argumentation, in order to prove that what I object to is really a corruption of genuine Christianity, and no part of the original scheme." *The Corruptions* took the historical approach of Priestley's first great book, on electricity, and played the tape backward: instead of a historical narrative of ever-increasing knowledge, it was a tale of ever-increasing obfuscation and error. To climb the mountain of Christian understanding, you had to go back to the very origins of the story.

For some of Priestley's peers, a work like *The Corruptions* presented a strange conundrum: how could it not occur to a radical materialist like Priestley that the very concept of God itself—whether it be Unitary, or Trinitarian, or Pagan—was the ultimate instance of supernatural thinking, the core distortion at the heart of most of the world's religions? (When he met with Lavoisier and other *philosophes* in Paris in 1774, they were startled to find that such an accomplished scientist was also a man of faith.) In *The Corruptions*, Priestley spends dozens of pages marshaling evidence showing how fifth-century theologians concocted the idea that God and Jesus were one, breaking from the original narrative that God had merely created Jesus to be a *human* messenger of his Word. But to someone existing outside the belief structure of Christianity—and even more, someone existing outside all organized religions—the two narratives would both seem to be in dire need of empirical evidence. If you don't believe in God, it's just as implausible to suggest that Jesus was a man created by God as it is to say that Jesus *was* God.

The concept of the nonexistence of the Christian God seems to have been a thought that Priestley was incapable of fully confronting. To a true atheist, the nonexistence of God defines the very edges of Priestley's intellectual map, the point beyond which he was unwilling to venture, the theological equivalent of phlogiston. But for a contemporary person of faith, the story reads differently: a religious man forced to alter and reinvent his beliefs—and challenge the orthodoxies of the day—in the light of science and history, who was nonetheless determined to keep the core alive. Priestley was a heretic of the first order who nonetheless possessed an unshakable faith. He seems to have been baffled by his intellectual peers who had made the leap into atheism. Priestley found it "lamentable" that a man of Ben Franklin's "good character and great influence should be an unbeliever in Christianity." But he attributed Franklin's non-belief to a lack of proper study. He wrote in his memoirs: "To me, [Franklin] acknowledged that he had not given so much attention as [he] ought to have done to the evidences of Christianity, and desired me to recommend to him a few treatises on the subject." Priestley loaded him up with some Hartley and a few volumes from his own *Institutes*, "but the American war breaking out soon after, I do not believe that he ever found himself sufficiently at leisure for the discussion."

But if Priestley failed to bring Franklin back among the faithful, he would have much better luck with another founding father with strong deist tendencies: Thomas Jefferson. Ironically, it was *The Corruptions* itself—a work devoted to dismantling so many of the central values of modern Christianity—that finally gave Jefferson enough philosophical support to call himself a Christian again.

It is not known when Jefferson first read *The Corruptions*. The edition in his famously comprehensive library—the seed of the modern Library of Congress—dates from 1793, and it seems likely that Jefferson read the book somewhere in that period, during his short-lived retirement back to the cerebral life of Monticello after serving as Washington's secretary of state. (Given Jefferson's exhaustion with the increasingly petty and partisan bickering of the new nation's leadership, reading a 300-page treatise on radical theological history would likely have been pure escapism for him.) What we know for certain is that the book made an indelible impression on Jefferson. Twenty years later, he would write to John Adams: "I have read [Priestley's] Corruptions of Christianity, and Early Opinions of Jesus, over and over again; and I rest on them ... as the basis of my own faith. These writings have never been answered." Shortly after assuming office as president in 1801, Jefferson wrote a much scrutinized letter to Benjamin Rush, defending his Christian beliefs against the many attacks he suffered during the contest with John Adams: "I am a Christian, in the only sense in which he wished anyone to be: sincerely attached to his doctrines in preference to all others, ascribing himself every human excellence and believing he never claimed any other." He prefaced this momentous declaration with a direct reference to Priestley: "To the corruptions of Christianity I am indeed opposed, but not to the genuine precepts of Jesus himself." When he constructed the legendary Jefferson Bible—a mash-up of original scripture, in which Jefferson selectively edited out all the references to Jesus's divinity and other supernatural elements—he was following a blueprint that Priestley had first drawn up in 1782.

Why did *The Corruptions* have such a profound effect on Jefferson? In one crucial sense, the book helped him find a way out of a bind he had struggled with for years. Alone with Franklin, Jefferson was the founder who most clearly embodied the Age of Reason, and while he never reached Franklin's level of accomplishment as a practicing scientist, he had a great passion for natural philosophy. (And for natural history, as evidenced by rich botanical and geological studies in the only book he ever published, *Notes on the State of Virginia*.) For most of his adult life, he had struggled to reconcile that faith in reason with a faith in the Christian God—or indeed in any organized religion. ("I am a sect by myself, as far as I know," he wrote.) His deistic tendencies were well known, and deployed against him throughout his public life, particularly in the campaign against Adams in 1800–01. (The pamphleteers also accused him of fathering children with one of his slaves, a charge that we now know was true—yet another reminder that political mudslinging is as old as the Union itself.) Jefferson's Enlightenment sensibilities made it difficult for him to keep his Christian faith alive, but the political realities of the day made it equally difficult for him to renounce Christ altogether. Priestley's *Corruptions* showed him the way out. Christianity was not the problem; it was the warped, counterfeit version that had evolved over the centuries that he could not subscribe to. Thanks to Priestley, he could be a Christian again in good faith—indeed, his Christianity would be purer, more elemental, than that of believers who clung to the supernatural trappings of modern sects.

The *Corruptions* resonated so strongly with Jefferson for another, more poetic, reason. The narrative structure of Priestley's story—an original state of purity and grace and moral cohesion, subsequently contaminated by schemers, charlatans, and elites—had a recurring presence in Jefferson's worldview, most notably in what he called the "ancient

Whig principles" of the original Anglo-Saxon culture: "a long-lost time and place," as the historian Joseph Ellis describes it, "where men had lived together in perfect harmony without coercive laws or predatory rulers," viciously warped by generations of kings, priests, and urban financiers into the loathsome form of eighteenth-century England. Just as Monticello—and the agrarian lifestyle in general—offered Jefferson a way back to that promised land, Priestley's *Corruptions* pointed the way to an equivalently pristine origin point where he could "rest his faith" without compromise. No doubt Jefferson admired Priestley's scholarship and his nimble close readings. But *The Corruptions* also made such an indelible impression for a simpler reason: it was the kind of story that Jefferson liked to hear.

IN THE BRITAIN of 1782, however, the story of Priestley's *Corruptions* did not fall on such sympathetic ears. Unsurprisingly to everyone but perhaps Priestley himself, *The Corruptions* stirred up an intense backlash after its publication, led by the Archdeacon Samuel Horsley, who denounced the work as an "extraordinary attempt . . . to unsettle the faith, and break up the constitution, of every ecclesiastical establishment." Horsley, who shared Priestley's passion for science, had nothing but contempt for dissenters, and he took the publication of *The Corruptions* as an opportunity to challenge Priestley's general reputation, dismissing his Copley Medal as the result of a few "lucky" discoveries and extracting a long list of mangled quotations and circular arguments from Priestley's oeuvre. Priestley soon got drawn into this "rude attack"—as he called it—and the melee continued for several years.

Partly as a response to his critics, Priestley gave a sermon in 1785 that was subsequently published as a pamphlet called *The Importance and Extent of Free Enquiry*. This would be the second tremor that presaged the eventual quake. It included a rallying cry for the Unitarian movement: "Let us not, therefore, be discouraged, though for the present we should see no great number of churches professedly Unitarian," he wrote. "It is sufficiently evident that Unitarian principles are gaining ground every day." But Priestley recognized that something more than mere optimism was warranted here, given the political realities of the time. He needed a metaphor for cultural change that could account for long periods of relative stability, followed by sudden revolutions. At first, he turned to the seasonal energy flows of agriculture: "We are now sowing the seeds which the cold of winter may prevent from sprouting, but which a genial spring will make to shoot and grow up; so that the field which to-day appears perfectly naked and barren, may to-morrow be all green, and promise an abundant harvest." But the steady cycle of the seasons were perhaps too tame for this story; in the next sentence, Priestley reached for a more intense energy metaphor: the earthquake, volcano, or vortex. "The present silent propagation of truth," he wrote, "may even be compared to those causes in nature, which lie dormant for a time, but which, in proper circumstances, act with the greatest violence."

And then Priestley took one fateful step up the ladder: from the energy flows of sunlight and vegetation, to the sudden eruptions of natural disasters, all the way up to the manmade explosions of war:

> We are, as it were, laying gunpowder, grain by grain, under the old building of error and superstition, which a single spark may hereafter inflame, so as to produce an instantaneous explosion; in consequence of which that edifice, the erection of which has been the work of ages, may be overturned in a moment, and so effectually as that the same foundation can never be built upon again.

Sparks, flames, explosions: the capture and release of energy remained central to Priestley's career, even rhetorically. Josiah Wedgwood had advised Priestley to cut the "gunpowder" line, predicting, accurately enough, that the phrase would incite too much controversy and distract from the sermon's otherwise more reasonable message. It didn't help matters that Priestley delivered the sermon on November 5, the anniversary of Guy Fawkes's failed 1605 attempt to blow up Parliament to protest the anti-Catholic laws of the day. The address became forever known as the "Gunpowder Sermon," and among his enemies, Priestley's nickname became "Gunpowder Joe." Incensed by the remarks, Horsley launched a new attack on Priestley and Lindsey, arguing that both had neglected to sign documents necessary for the legal recognition of their meetinghouses. Priestley, in particular, posed a direct threat to the state, Horsley argued; in one ominous sentence, he encouraged the "trade of the good town of Birmingham . . . to nip Dr Priestley's goodly projects in the bud."

The final of the three tremors originated across the Channel, in the opening act of the French Revolution, which Priestley and most of the Lunar Society greeted with intense interest, seeing it as the logical continuation of the enlightened progress that had begun with the American uprising the previous decade. Priestley's old friend from the Honest Whigs, Richard Price, delivered a sermon that enthusiastically linked the two revolutions. Priestley immediately wrote to congratulate Price, celebrating "the liberty, both of that country and America, and of course of all those other countries that, it is to be hoped, will follow their example." Shortly thereafter, Edmund Burke penned his classic *Reflections on the Revolution in France, and on the Proceeding in Certain Societies in London Relative to That Event.* It was a direct rebuttal to Price and Priestley, and to the radical Whig groups that had embraced the news from France with such enthusiasm. Burke dismissed the group as a pack of naïve idealists, "unacquainted with the world in which they are so fond of meddling, and inexperienced in all its affairs, on which they pronounce with such confidence." He playfully alluded to Priestley's *Observations on Air* and his soda-water invention in one oft-quoted line:

The wild *gas*, the fixed air, is plainly broke loose: but we
ought to suspend our judgment until the first effervescence

is a little subsided, till the liquor is cleared, and until we see something deeper than the agitation of a troubled and frothy surface.

Events would ultimately prove Burke right, at least about the difficulty of making sense of the Revolution's initial froth. But Priestley reacted with unusual hostility in a pamphlet published shortly after Price's. "Your whole book, Sir, is little else than a vehicle for the same poison," he wrote, "inculcating, but inconsistently enough, a respect for princes, independent of their being originally the choice of the people as if they had some natural and indefeasible right to reign over us, they being born to command, and we to obey."

To PRIESTLEY, ever the optimist, the controversies of the 1780s seemed like an indisputable sign of progress, both personal and societal. His ideas on religion and politics had reached the level of influence that his natural philosophy had attained during the Leeds years a decade before. That there was resistance to the grains of gunpowder being laid was inevitable; the central, undeniable point was that change was on the march. Yet almost all the core elements from this period of Priestley's life—the coal deposits, the new factory system, the empowered dissenting churches, the revolutions abroad—conspired to produce a kind of dialectical monster that would rise up to take its vengeance on everything that Priestley and his coterie stood for. This was the "Church and

King" movement, a reactionary band of largely working-class men, incited by the conservative elites, hostile to change in all its diverse forms: the dissenters undermining the Church of England; the industrial magnates that had destroyed the agrarian tranquillity of rural England; the aspiring regicides overseas who sought to put an end to all forms of monarchy, aided and abetted by their allies on British soil. With his mix of religious radicalism, Francophile tendencies, and membership in the Birmingham elite, Gunpowder Joe was the ultimate nemesis for the mobs of Church and King.

There were warning signs. A self-described "button burnisher" with the alias John Nott penned a public letter to Priestley that included this not-so-veiled threat: "Now, prithee Mr. Priestley, how would you like it yourself, if they were to send you word that they had laid trains of gunpowder under your house or meeting-house?" In January of 1790, three men attempted to break into his house at Fair Hill, firing a pistol at a maid through an open window after their presence was detected. Nott published a coyly ominous note shortly thereafter: "Don't you remember what a parlous taking you was in one Saturday, when one of our Birmingham gunners shot at a flight of sparrows in your garden thinking no harm."

But the "wild gas" of Church and King in Birmingham would not fully break free until July of 1791, when the newly formed Constitutional Society—which numbered Priestley among its members—announced plans for a dinner on Bastille Day, welcoming "any Friend to Freedom" to join them at the Royal Hotel on Temple Row. When a second advertisement appeared in the *Birmingham Gazette*, a separate, anonymous note ran alongside it, threatening to publish an "authentic list of all those who dine at the hotel" that night. It was signed "Vivant Rex et Regina." A succession of leaflets, handbills, and newspaper adverts rolled in over the next week, inciting tempers on all sides. The most incendiary was a veritable call to arms: "Whatever the *modern republicans* may imagine, or the *regicidal propounders of the rights of men* design, let us convince them there is enough loyalty in the majority of the inhabitants of this country, to *support* and *defend* their King." By the morning of the 14th, there were various rewards offered for any evidence of the authorship of several of the leaflets on either side. The Constitutional Society itself took out an advertisement in the *Gazette*, reaffirming its belief in the three estates of King, Lords, and Commons, without backing down entirely from their support of the French revolt: "Sensible themselves of the advantages of a Free Government, they rejoice in the extension of Liberty to their Neighbours, at the same time avowing, in the most explicit manner, their firm attachment to the Constitution of their own Country."

These last-minute concilitory gestures proved futile. In the long, diverse history of humans gathering together to celebrate over a meal, the Constitutional Society meeting on July 14 may well rank as the most politically explosive dinner party on record. And the irony of it is that the dinner very nearly didn't take place at all. At some point in the morning of the 14th, the society called off the event, but the hotel proprietor—perhaps concerned about losing a lucrative booking—suggested an alternative plan: they carry on with the dinner, but leave early, before the inevitable trouble started. Priestley, however, took the counsel of his friends and remained at Fair Hill.

At three o'clock, roughly eighty men arrived at the hotel, showered by slightly confused jeers of "No Popery!" from a small Church and King crowd that had gathered by the door. A local artist had created a sculpture for the occasion: a medallion of King George, framed by two obelisks that symbolized "British liberty in its present enjoyment," the second "Gallic liberty breaking the bands of Despotism." The first toast of the dinner—somewhat defensively—was to the king and the Constitution, but it was followed shortly by glasses raised to the French National Assembly, among other Whig causes. By five the event was over, and the Constitutional Society disbanded rapidly under attack from the protestors, who had replaced their slogans with rocks. It seemed, at first, that a more serious conflagration had been avoided, until around eight p.m., when a much larger group of protestors emptied out of the pubs and arrived at the doorstep of the Royal Hotel, thinking that the dinner was due to end at that hour.

The discovery that they had missed their regicidal foes by three hours appeared to pique the mob's anger; the windows at the front of the hotel were smashed before a cry went up to move on to the New Meeting House. Armed with crowbars and bludgeons, the mob reduced the entire structure to a smoldering shell within an hour, the gates, doors, pews, and books dragged from inside the church and piled up into an enormous bonfire on the front steps. Another faction marched to the Old Meeting House and lit another fire in the small cemetery next to the church; a blaze inside the church grew so intense that the roof eventually collapsed.

As the mob destroyed the Birmingham meetinghouses, at Fair Hill, on the outskirts of town, Joseph and Mary Priestley were engaged in a quiet game of backgammon, entirely oblivious of the chaos only a mile away. At ten p.m., Priestley's friend Samuel Ryland stormed into the house, having raced from Birmingham by chaise, with news of the riots. The meetinghouses were destroyed, he reported to the startled couple, and the mob was now on the march. Priestley and Fair Hill were its next target.

WHAT IS THE INTERNAL chemistry of a mob? They are the waterspouts of social history: rare but powerful expressions of force, capable of erupting into existence with little warning, and dissipating just as quickly. Tellingly, mob behavior inevitably gravitates toward displays of intense energy transfer: the collective strength of a hundred enraged men pulling a building apart and unleashing the destructive, oxidizing force of combustion. In this sense, there is something truly primitive about a mob, not just in the sense of men—and it is almost always men—regressing to a precivilized state of unorganized violence, but also in the obsession with that most primitive form of energy release: fire itself.

Historians have long debated whether the mobs of the Birmingham Riots were a spontaneous expression of rage— leaderless and self-organizing—or whether they had been deliberately stoked by outsiders. (The evidence suggests that it was a little of both.) But by the time Church and King protestors arrived at Fair Hill, the madness of the crowd was beyond the direct control of the original ringleaders, whoever they were. A later report claimed that the insurgents had brought an immense gridiron to Fair Hill, "where they said they meant to broil an anti-constitutional philosopher, by the blaze of his own writings, and light the fire with the *Rights of Man*."

Mary and Joseph had retreated to Showell Green, the estate of William Russell, a close friend of the Priestleys and a prominent Birmingham merchant. They could see the Meeting House fires burning in the summer night sky as they rode through the dark. Russell himself left his family behind with Priestley's and rode his horse toward the Birmingham center before friends forced him to retreat to Fair Hill. There, he joined the twenty-year-old William Priestley, who, along with a handful of servants, had stayed behind to protect the house and salvage the most valuable books and manuscripts. When the mob arrived at the Fair Hill gates, Russell urged them to disband, and for a few minutes his words seemed to have a pacifying effect. But before long, shouts of "Stone him! Stone him!" erupted from the crowd, and Russell and

Burning of Dr. Priestley's House at Fairhill.*

William Priestley were forced to return to Showell Green, with the somber news that the mob had taken control of Fair Hill. Fearing that the mob would venture to the Russells' next, the two families set off near midnight for the home of Thomas Hawkes, a half mile away. They left a barrel of ale on the lawn as a peace offering to the rioters.

A small farce ensued on Fair Hill, as the mob was apparently incapable of starting a proper fire, their arson skills no doubt impaired by the gin and beer, along with the wine they'd discovered in Priestley's cellar. (According to Priestley's own somewhat derisive account, they had even attempted to extract flame from the electrical machine he had used to entertain the children in the upstairs library.) But eventually the mob stumbled across Priestley's laboratory, which had been built at a distance from the main house and was amply stocked with tools for combustion. Within a matter of hours, Fair Hill was gone: the library where Priestley had performed magic lantern shows for the Lunar children, the drawing room where Mary and Joseph had played their backgammon, thousands of manuscript pages documenting decades of Priestley's investigations, the laboratory he had lovingly built for himself, along with that unique collection of tools that his Birmingham friends had crafted for him over the years. All of it had been lost to the fire.

At three o'clock Russell ventured to Fair Hill and found the rioters dispersed across the lawn, most of them in a drunken slumber amid the smoldering rubble. He returned to the Hawkes house and informed the families that it was likely safe to return to Showell Green. The Russell's daughter, Mary, later recalled:

> Accordingly we set off, and never shall I forget the joy with which I entered our own gates once more. . . . A room was prepared for the Doctor and Mrs P. We all looked and felt our gratitude; but the Doctor appeared the happiest among us. Just as he was going to rest, expressing his thankfulness in being permitted to lie down again in peace and comfort, my father returned from Fair Hill with the intelligence that they were collecting again, and their threats were more violent than ever, that they swore to find Dr. P and take his life.

With dawn about to break, and the prospect of the riots growing in intensity, the group of refugees realized that they had no clear path of retreat. Eventually it was decided to leave the Birmingham environs altogether, and head to the outskirts of Dudley, almost ten miles away. After a day in Dudley, they traveled to Worcester, hoping to catch the postal carriage to London, but got lost in the rural darkness and spent a bleary night wandering across the commons between Bridgenorth and Heath Forge. Eventually they reunited with Ryland, who offered Priestley his wig and cloak as a disguise. Priestley declined. By the morning of the 18th, they had made it to Reverend Lindsey's in the Strand. For several weeks, Priestley lived underground in London, an exile in his own country, not daring to show his face in public, just as Franklin had done fifteen years before.

Back in Birmingham, the inferno was slowly dying out. At the king's request, three troops of Dragoons had arrived on the 17th to subdue the riot. (Many thought the response time was suspiciously slow.) By the time it was over, more than a dozen homes and churches had been razed by the mob, including Russell's and Ryland's. Dozens of rioters lost their lives, including ten at Ryland's home who were both buried and burned alive when a flaming roof collapsed on them in the cellar.

WHAT HAPPENS IN the mind of one of the world's great optimists when his work inspires his fellow countrymen to rise up and destroy his home and the tools of his trade? On the exterior, Priestley was by all accounts a portrait of remarkable self-composure, given the devastation and the still-imminent threat to his life. Mary Russell would later describe his demeanor on that dark night:

> Undaunted he heard the blows which were destroying the house and laboratory that contained all his valuable and rare apparatus and their effects, which it had been the business of his life to collect and use. . . . [H]e, tranquil and serene, walked up and down the road with a firm yet gentle pace that evinced his entire self-possession, and a complete self-satisfaction and consciousness which rendered him thus firm and resigned under the unjust and cruel persecution of his enemies. . . . Not one hasty or impatient expression, not one look expressive of murmur or complaint, not one tear or sigh escaped him; resignation and a conscious innocence and virtue seemed to subdue all these feelings of humanity.

Still, it seems hard to imagine that Priestley maintained that kind of calm in the face of such brutality. His "Letter to the Inhabitants of Birmingham, Following the Riots of 14 July 1791" took a more aggressive tone. Addressing his "late townsmen and neighbors," he began with a protestation of innocence that was perhaps a bit too strong, given the militant rhetoric he had indulged in during the preceding decade. "After living with you 11 years," he wrote "in which you had uniform experience of my peaceful behaviour, in my attention to the quiet studies of my profession, and those of philosophy, I was far from expecting the injuries which I and my friends have lately received from you." A natural philosopher to the end, Priestley then moved quickly to the core injustice—not the threat to his life, his family, his home, but the loss of his gear:

> You have destroyed the most truly valuable and useful apparatus of philosophical instruments that perhaps any individual, in this country or any other, was ever possessed of, in my use of which I annually spent large sums of money with no pecuniary view whatever but only in the advancement of science, for the benefit of my country and of mankind.

By the end of the letter he returned to the central insight that had launched his decade at Fair Hill and had sent him down the path to the Birmingham Riots—the corruptions of original Christian values: "We are better instructed in the mild and forbearing spirit of Christianity than ever to think of recourse to violence—and can you think that such conduct as yours [offers] any recommendation of your religious principles in preference to ours?"

Between the tranquillity of Mary Russell's account, and the firebrand rebuttal of Priestley's letter, the most revealing look at Priestley's inner state—and the emotional weight of his loss—comes in a letter he wrote to a friend some time after the riots: "I shall be obliged to you," he wrote, "if you will mention my situation to any of your friends whose laboratories are furnished, and who may have anything to spare to set up a broken philosopher."

THE RIOTS SENT a shock wave through British society, though the establishment generally adopted a blasé attitude that suggested Priestley and his ilk had it coming to them. The king's order to send the Dragoons had included this withering remark: "I cannot but feel better pleased that Priestley is the sufferer for the doctrines he and his party have instilled, and that the people see them in their true light." The *Times* even ran an entirely scurrilous report of the dinner, which falsely placed Priestley at the event, and quoted him raising his glass with a toast to "The King's head on a platter."

Other voices were more sympathetic. The Lunar Society, of course, rallied to Priestley's side—Darwin called the riots a "disgrace to Mankind"—though they did gently warn him to moderate some of his public views, to protect himself and their circle of friends. Dissenting churches around the country expressed support for their spiritual and political guiding light. Vice President John Adams sent his letter of support from America, comparing Priestley's persecution to that of Socrates. Perhaps the most touching note of solidarity came from across the Channel, in a statement issued by the French Academy of Sciences, and likely penned by Lavoisier himself. "As a citizen, you belong to England, and it is to her to atone for your losses: as a Scholar and as a Philosopher you belong to the entire world," they wrote. And in a fitting gesture, they offered to help reconstruct Priestley's ruined laboratory: "We . . . vow to restore to you the instruments which you have employed so usefully in our instruction. . . . What more important service can we render to science than to place in your hands the instruments necessary for its cultivation?"

Aided by his allies in the Lunar Society, Priestley waged a long campaign to be compensated by the state for his losses. He settled in Hackney, taking up ministerial duties at Richard Price's old congregation. (Price had died in April of 1791, leaving Priestley alone as England's public enemy number one.) Though Priestley found that most members of the Royal Society shunned him "on account of [his] religious and political views," over time, some semblance of normalcy would return to his life, with Joseph publishing again and preaching his Sunday sermons. But the Priestleys would never feel fully at home in England again. In August of 1792, the French legislative assembly endowed Priestley with an honorary citizenship, which triggered a whole new round of cartoons and angry pamphlets deriding Priestley for his traitorous ways. In October 1793, Joseph Jr. emigrated to Pennsylvania and sent back word of a promising land settlement between the two branches of the Susquehanna, a hundred and fifty miles northwest of Philadelphia. By the spring of 1794, Joseph and Mary Priestley had set sail on the *Samson* for America.

On the news of Priestley's departure, a twenty-one-year-old Samuel Taylor Coleridge penned these lines:

*Lo! Priestley there, Patriot, and Saint and Sage!*
*Him, full of years, from his lov'd native land,*
*Statesmen blood-stain'd, and Priests idolatrous,*
*By dark lies maddening the blind multitude,*
*Drove with vain hate. Calm, pitying, he retired,*
*And mused expectant on these promised years.*

Dr PRIESTLEY.

## CHAPTER FIVE

# A Comet in the System

*February 1804*
*Northumberland, Pennsylvania*

THE *SAMSON* HIT HEAVY RAINS AND impenetrable fog as she neared the northeast coast of America but eventually found her way to Sandy Hook on June 1, where she anchored for a few days, waiting for a pilot to steer her through the channels of New York Harbor. On June 4, Joseph Jr. hired a boat to greet his parents and bring them back discreetly to the Battery. Despite his efforts, word soon spread of Priestley's arrival, and by the next morning an august procession of luminaries arrived to pay their respects, starting with Governor Clinton. Vice President Adams had left New York the day before Priestley's arrival, but he had left behind a note encouraging Priestley to settle in Boston, and promising that he would be very well received there. This began a bit of an internal fight among the American founders over where Priestley should settle, and reading the letters now, one senses that the debate was ultimately a proxy for a larger dispute over where the intellectual center of the new country lay.

The New York papers printed solemn declarations of support and welcome from scientific and religious societies around the city. The *American Daily Advertiser* pronounced: "The name of Joseph Priestley will be long remembered among all enlightened people. . . . His persecutions in England have presented to him the American Republic as a safe and honourable retreat in his declining years: and his arrival in this City calls upon us to testify our respect and esteem for a man whose whole life has been devoted to the sacred duty of diffusing knowledge and happiness among nations."

Priestley wrote back to Lindsey to say that the reception had been "too flattering." The accolades would continue several weeks later, when the Priestleys moved on to Philadelphia, where they were met with "the most flattering attentions from all persons of note," according to Priestley's account. He enjoyed tea several times with President Washington, and cemented a friendship with Benjamin Rush, the physician and political theorist, whose writing career came closest to matching Priestley's in the diversity of its subject matter. (With Franklin gone, Rush would become the American whom Priestley found the most "congenial.") Priestley's appearance in Philadelphia precipitated a formal address from the astronomer David Rittenhouse, one of Jefferson's intellectual mentors and president of the American Philosophical Society, of which Priestley had been a long-distance member since 1785. When word of Priestley's arrival trick-led down to Jefferson, who was at Monticello, enjoying his first hiatus from political life, he wrote Rittenhouse envi-ously: "If I had but Fortunatus's wishing-cap, to seat myself sometimes by your fireside, and to pay a visit to Dr. Priestley, I would be contented; his writings evince that he must be a fund of instruction in conversation, and his character an object of attachment and veneration."

The sense of gravitas that attended Priestley's emigration seems somehow fitting to us now, not just because of his individual accomplishments, but also because Priestley was inaugurating what would become one of the most honorable traditions of the American experience. He was the first great scientist-exile to seek safe harbor in America after being per-secuted for his religious and political beliefs at home. Albert Einstein, Edward Teller, Xiao Qiang—they would all follow in Priestley's footsteps.

Priestley initially embraced asylum in America with his typical enthusiasm: he was, at long last, in his own element, surrounded by intellectual peers who also happened to be, amazingly enough, the political establishment. "Whether it be the effect of general liberty, or some other cause," he wrote to Lindsey in June, "I find many more clever men, men capa-ble of conversing with propriety and fluency on all subjects relating to government, than I have met with any where in England. I have seen many of the members of Congress on their return from it, and, without exception, they seem to be men of first-rate ability."

The story up to this point, one month into his emigration, seems to provide an almost irresistible narrative arc: persecuted scientist and priest leaves behind the repressive fossils of the Old World to thrive among the statesmen-scholars of the new republic—a country whose creation he himself had supported from afar. Alas, the story is not quite so neat, though it is far more interesting, largely for what it reveals about the birthing pains of the new nation. Priestley may have imagined that he was escaping England to retreat back to the private pursuits of natural philosophy, but once again he would find himself on the main stage of history, this time in his adopted homeland.

THE FIRST TURN OF EVENTS that sent the Priestleys off course was a disappointment in the real estate market. The land that Joseph Jr. had scouted as a site for a large settlement of like-minded émigrés turned out to be less desirable than originally thought, and by the fall of 1794 the entire scheme had collapsed, upsetting Priestley greatly. The University of Pennsylvania offered him the position of chair of the chemistry department, and for a time the Priestleys debated staying in the city. But Philadelphia at that point was pestilent and overcrowded, ravaged by yellow-fever epidemics, and almost entirely lacking green space. Mary Priestley had always preferred rural to urban life, and in recent years she had begun to suffer from ominous episodes that involved spitting up blood for several days. So the couple decided to stick with their original plan to settle in central Pennsylvania—in Northumberland—this time on a less ambitious scale—accompanied by their children and a few friends who settled nearby, including Thomas Cooper, the chemist and political agitator who had also emigrated from England in 1794. For the first time, the couple planned a house together, to be built on eleven acres outside the small town, and Priestley once again reconstructed his laboratory, his fragile instruments damaged by "injudicious packing" on the trip from Philadelphia. Mary was overjoyed with their new life: "I am happy and thankful to meet with so sweet a situation and so peaceful a retreat as this place I now write from," she wrote back to England.

Joseph grumbled about the effect the sluggishness of the postal system was having on his work. In London, Birmingham, and Leeds, information had traveled on the scale of hours or days. Communicating with the Honest Whigs or the Lunar Society via mail back home had been a conversational experience: you could make plans, or banter, jot off quick observations, swap half-formed ideas, at that accelerated rhythm. But the lag time just between Northumberland and Philadelphia was often a matter of weeks, and sending a message all the way to London took an entire season at least. This meant a detachment from world news as much as it did from personal connection. "I could now give a great deal for a complete set of the Morning Chronicle," Priestley wrote to Lindsey, "or any tolerable English newspaper tho ever so old. I hope Mr. Belsham will send me the Cambridge Papers.

They would amuse me much. We have only poor extracts in the Philadelphia papers." Priestley had experienced once before what it was like to be separated from his tools, but in the move to Northumberland he felt for the first time the pain of separation from his information network, or at least of seeing its transfer rates decline by an order of magnitude or two. He wrote several appeals to Adams, imploring him to help establish a regular stagecoach to Northumberland: "Could we have a Coach . . . to carry parcells, and passengers, as well as letters, it would be a great convenience and benefit to the country, and in time would pay for any reasonable expence attending it," Priestley wrote. "We sometimes talk of petitioning the legislature on the subject. Could you give us any assistance in the business, you would confer a great obligation on one who was so much interested in the conveyance of letters and small parcells."

If Priestley's natural philosophy suffered from his being unplugged from his usual network, an even greater blow was dealt by the vicissitudes of health and disease. The first strike came in December of 1795, when their son Harry died somewhat suddenly, after battling a wave of fevers. Both Mary and Joseph were devastated, but then Mary's health took a turn for the worse, and they spent most of the winter recovering, emotionally and physically, from the loss of Harry. In late February, Priestley returned to Philadelphia for several months, trying to re-create the annual migrations to the metropolis that had so energized him in England. (He stayed with William and Mary Russell, his short-lived protectors during the

Birmingham Riots, who had themselves emigrated, in a tortuous, prolonged voyage the year before.) Priestley quickly returned to the pulpit, delivering sermons at the Universalist Church on Lombard Street.

Priestley's first addresses were well attended: a flock of luminaries from the Revolutionary War came out to hear the legendary Gunpowder Joe talk. Vice President Adams attended, and reported backed to Abigail Adams that the sermons—titled "Discourses on the Evidences of Divine Revelation"—were "learned, ingenious, and useful." But even Adams was more cautious in endorsing such a controversial figure. "The [Discourses] will be printed, and He says dedicated to me," he wrote to Abigail. "Dont tell this secret though, for no other being knows it. It will get me the Character of an Heretick I fear. I presume however, that dedicating a Book to a Man, will not imply that he approves every Thing in it."

Over time the novelty of the famous radical's sermons appears to have worn off and the number of people showing up on Lombard Street began to dwindle. Adams in particular grew disenchanted with Priestley. Some scholars have attributed the growing distance between the two men as a case of political pragmatism: Priestley was probably the most controversial religious figure of the age, and Adams was running for president. It was one thing to be a follower of a great champion of the American cause, but it was quite another to throw your lot in with a minister who considered half of modern Christianity to be a bunch of Pagan hocus-pocus. Here again Priestley lies at the origin point of another venerable—if not altogether meritorious—American tradition: aspiring politicians distancing themselves from their controversial religious advisers during the campaign season.

But Adams may have soured on Priestley for another reason: Priestley's increasingly millenarian tendencies. At one of their last meetings, Priestley and Adams had breakfast alone together, and Adams began inquiring about his friend's thoughts on the Reign of Terror in France. Priestley was typically sanguine. The Revolution was "opening a new era in the world and presenting a near view of the millennium." Adams thought Priestley seemed a notch too blithe about such a volatile and uncharted situation, and pressed him to explain how he could be so sure of France's democratic prospects. The answer startled the vice president. "My opinion is founded altogether upon revelation and the prophecies," Priestley explained. "I take it that the ten horns of the great Beast in revelations, mean the ten crowned heads of Europe: and that the execution of the king of France is the falling off of the first of those horns; and that the nine monarchies of Europe will fall one after another in the same way."

Priestley published almost none of his millennial speculations; reading the Book of Revelation for political prophecies seems to have been something of a private hobby for the old man, one that his friends often chided him about. The epic explosions of the French Revolution and Priestley's exile into the wilderness of Pennsylvania intensified these thoughts, and so the great believer in rational Christianity would spend some not insignificant portion of his last years mulling over the horns of the Great Beast. Adams was baffled at the seeming contradiction: a man dedicated to excising every last hint of mysticism from the New Testament, but who nonetheless happily based his interpretation of contemporary political events on hallucinatory visions from Revelation. Priestley's erratic behavior over breakfast might have cost him some regard in the eyes of John Adams, but in several years' time, his apocalyptic musings would indirectly come to his aid and help protect him from a potential threat to his freedom nearly as severe as that of the Birmingham Riots.

Before Priestley returned to the center stage of political controversy, he had his own private tragedy to endure. Upon his return from Philadelphia in the middle of 1796 he found Mary's condition much worsened. By September, she was dead. The intellectual isolation from the Lunar Men that Priestley felt would be nothing compared to this loss. "The death of Harry affected her much, and it has hardly ever been out of my mind, tho it is near 9 months since he died; but this is a much heavier stroke," he wrote in a letter. "It has been a happy union to me for more than thirty years, in which I have had no care about anything in the world, so that, without any anxiety I have been able to give all my time to my own pursuits. I always said I was only a lodger in her house." Part of what made Mary's death so painful to Priestley was the fact that she, more than any of the Northumberland settlers, had genuinely embraced life on the frontier. "She had taken much pleasure in planning our new house," he wrote mournfully, "and now that it is advancing apace, and promises to be every thing that she wished it to be, she goes to occupy another." A few months after Mary's death, he wrote to her brother, John Wilkinson: "Having always been very domestic, reading and writing with my wife sitting near me, and often reading to her, I miss her every where."

With Mary's death, a long-standing dream of Priestley's finally expired as well: there would be no great settlement in Northumberland, no Lunar Society on the Susquehannah—in fact, there would be very little companionship at all. He was too frail to move again, and with Mary gone, he was dependent on the care of Joseph Jr., who had begun to make a sensible life for himself as a farmer. He had held out hope that his old student Benjamin Vaughan or the Russells would settle in Northumberland, but Vaughan ended up in Maine, and the Russells were appalled by the rustic lifestyle Priestley had adopted. (His house, Thomas Russell wrote, was "a mere hut in comparison with the one they lived in formerly.") Lonely in the woods, isolated from civilization and the news of the world, Priestley went so far as to initiate a few halfhearted plans to emigrate again, this time to France, but it never amounted to anything.

Yet just as the exhilaration of Priestley's arrival in America would prove ephemeral, so would the gloom of 1796 ultimately pass, in part because of the one undeniably positive development during this otherwise stormy period in Priestley's life. He became close friends with Thomas Jefferson.

JEFFERSON HAD RETURNED to Philadelphia on March 2, 1797, after a four-year retirement from politics at Monticello. He had a busy week on his return. The day after his arrival he was installed as president of the American Philosophical Society, succeeding David Rittenhouse, who had died several months before. And on March 4, Jefferson was sworn in as vice president of the United States, serving under the new president John Adams. Priestley was in Philadelphia that spring, and, his relations with Adams being generally strained, he began spending a great deal of time with the new vice president.

This small, seemingly private shift in the personal connections between these three men—one friendship fading, to be replaced by another—ended up having significant political consequences for the early years of the Republic. The complex dynamic between Jefferson, Adams, and Priestley would continue to play out for more than twenty years, outliving Priestley by more than a decade.

To choose between Jefferson and Adams in 1798 was, in effect, to choose between the two emerging political parties to which each man had become attached: the largely agrarian Republicans, led by Jefferson and Madison, opponents of centralized political and economic power; or the urban, centralized Federalists, led by Alexander Hamilton and, somewhat fitfully, by Adams. Given the geopolitics of the day, it was also a choice between France and England, with Jefferson and his group still enthralled by the French Revolution and Hamilton aligned with London's economic power.

Choosing sides had an even more profound implication as well, one that no longer applies in our modern political world. To align yourself with either party in 1798 was to endorse the whole concept of different political parties, which was then a new and fiercely contested development on the American political scene, following the ostensibly unified front of the first two Washington administrations. Adams had long resisted the idea that opposing parties were an inevitable development. (They were to be "dreaded as the greatest political evil under our Constitution," he wrote.) Priestley himself had warned against the emergence of political factions in his private correspondence, shortly after his arrival in America. But by 1798, perhaps feeling spurned by his falling-out with Adams, Priestley had firmly thrown his lot in with the Republicans. A few weeks before Jefferson arrived in Philadelphia, Priestley wrote his first openly political tract since the Birmingham Letters, published in the pro-Jefferson Philadelphia newspaper, the *Aurora*. The main thrust of the argument was a protest against building up a national army and navy prepared for war with France, when America would be much better served by retreating from costly overseas entanglements at this fragile stage in its development. True to form, Priestley thought the war trust would be better spent on libraries and laboratories, "of which all the universities and colleges of this country are most disgracefully destitute." England and France would have much to fear from an American air pump, if the current administration would only see fit to endow its schools with the technology.

Priestley's "Maxims on Political Arithmetic" were published anonymously, but word soon leaked that Gunpowder Joe had begun laying grains under the Federalist party, just as he had done to the Church of England back home. Priestley's Northumberland neighbor Thomas Cooper began publishing even more strident attacks on the Federalists in the *Northumberland Gazette*, which Cooper edited, building something of a Republican stronghold in the wilds of Pennsylvania. Priestley engaged in a long correspondence with a Unitarian minister and congressman named George Thatcher, in which he stridently criticized Adams, calling the new president "unstatesmanlike" in his war-mongering with France. Before long, Priestley had become a popular target for the cartoonists and pamphleteers of the day, led by another British expatriate, one William Cobbett, who had been publishing screeds—under the pen name Peter Porcupine—against Priestley since the first days of his arrival in 1794. Priestley, Cobbett wrote, had entered America with the express aim of "disunit[ing] the people from their government, and . . . introduc[ing] the blessing of French anarchy."

With tensions rising across the country, Adams signed into law the notorious Alien and Sedition Acts in the summer of 1798, authorizing the state to deport any noncitizen "dangerous to the peace and safety of the United States," and to arrest anyone who published "false, scandalous, and malicious writing" about the government. It was the fledgling nation's first constitutional crisis. Jefferson, the sitting vice president, announced that signing of the acts into law meant that the government had "become more arbitrary, and ha[d] swallowed more of the public liberty than even that of England." To the opponents of the acts, the whole American experiment in democractic rule seemed at risk. Would the constitutional framework—still in its infancy—survive this challenge to its core values? Was Alien and Sedition the first step of an inevitable progression that would transform the young republic into a dictatorship, the same dismal trajectory that France was currently following across the Atlantic?

Concerned about his own situation under these new laws, Priestley wrote to Thatcher, urging him to keep their correspondence private, but word soon got back to Priestley that his criticisms of Adams were now well known in Philadelphia, including some details that seemed to have come from his exchange with the Maine congressman. He wrote back to Thatcher:

> I have not written to any person in Philadelphia besides yourself, and I am sure you would not *intentionally* expose me to danger. However, I will take care to *send no more, lest a worse thing come unto me. I* find I am at the mercy of one man, who, if he pleases, may, even without giving me a hearing, or a minute's warning, either confine me, or send me out of the country. This is not a pleasant situation.

The situation was about to get much worse. Unbeknownst to Priestley, a few weeks before the Alien and Sedition Acts passed, a packet of letters headed for Priestley was captured on board a Danish frigate and leaked to the British press. The letters had been penned by John Hurford Stone, a British radical whom Priestley had met in Price's congregation in Hackney. The correspondence addressed Priestley as a committed supporter of the French, and spoke rhapsodically of France's plans to invade England and complete its project of bringing the glories of liberty to all of Europe. Stone alluded to Priestley's plans to emigrate to France, and made dismissive comments about John Adams's leadership. It was an entirely one-sided conversation, but the undeniable impression on reading the letters was that Stone believed he was writing to a friend whose primary allegiances were to the Directoire Exécutif in Paris above all else.

On August 20, William Cobbett published the letters in their entirety, accompanied by scathing editorial commentary and a banner headline: "PRIESTLEY COMPLETELY DETECTED." The copy included a direct challenge to Adams: "If this discovery passes unnoticed by the government, it will operate as the greatest encouragement that its enemies have ever received; they will say, and justly too, that though the President is armed with power, he is afraid to make use of it, and that the Alien-Law is a mere bugbear."

Priestley was devastated by the uproar that followed. "I am considered as a citizen of France," he wrote back to an English friend, "and the rage against every thing relating to France and French principles as they say, is not to be described. It is even more violent than with you. This is a change that I was far from expecting when I came hither." For several months, he went silent, hoping that a retreat from the public sphere would calm the passions against him. (Here he may have learned a lesson from his immediate and confrontational public statement after the Birmingham Riots, which only served to fan the flames higher.) But by the new year he was back writing to Thatcher, more incensed than ever at the administration's violation of civil liberties: "It is clear to me," Priestley wrote, "that you have violated your constitution in several essential articles, and act upon maxims by which you may defeat the whole object of it." That congenital openness that had helped him so much in spreading the enlightenment of natural philosophy might end up getting him imprisoned or deported, but he was too old to change his ways:

> I may be doing wrong in writing so freely, and I have been desired to be cautious with respect to what I write to *you*. But I am not used to secrecy or caution, and I cannot adopt a new system of conduct now. There is no person in this country to whom I write on the subject of Politics besides yourself, nor do I recollect what I have written; but I do not care who sees what I write or knows what I think on any subject. You may, if you please, show all my letters to Mr. Adams himself.

Thomas Cooper also began fighting the administration more openly, publishing a series of fierce editorials in the *Northumberland Gazette*, starting in early 1799. (No doubt most of them emerged out of conversations with Priestley, given their isolated situation.) They ended with an address published in June that systematically laid out the case against the abuses of the Adams presidency: "I cannot help thinking that of late years, measures have been adopted and opinions sanctioned in this country, which have an evident tendency to stretch to the utmost the constitutional authority of our Executive, and to introduce the political evils of those European governments whose principles we have rejected." Adams's policies, in short, were exactly those "that a leader inclined to despotism might wish."

Cooper's address was reprinted in the *Philadelphia Aurora*, and circulated widely via handbills. (Allegedly, Priestley had assisted in their distribution.) For Secretary of State Thomas Pickering, it was the final straw. Pickering advised Adams that Priestley and Cooper were in clear violation of the Alien and Sedition Acts. Of Priestley, Pickering wrote, "What is of most consequence, and demonstrates the Doctor's want of decency, being an alien, his discontented and turbulent spirit that will never be quiet under the freest government on Earth, is his industry in getting Mr. Cooper's address printed in handbills and distributed." He noted ruefully that Cooper had naturalized himself as an American citizen: "I am sorry for it, for those who are desirous of maintaining our internal tranquillity must wish them both removed from the United

States." Adams agreed with Pickering's take on Cooper's address: "A meaner, a more artful, or a more malicious libel has not appeared. As far as it alludes to me, I despise it; but I have no doubt it is a libel against the whole government, and as such ought to be prosecuted." Later that year, Cooper was in fact arrested, and became one of the ten men successfully prosecuted under the Sedition Act.

But the question of Priestley was far more complicated. Four years before, after the senseless extremity of the Birmingham Riots, Adams had sent Priestley a letter comparing him to Socrates, a man of great wisdom persecuted by an unthinking establishment. Two years before, Adams had sat, attentive, in the first pews for Priestley's initial sermons in America. They had known each other personally for more than ten years. And yet now Priestley, a guest in this country, was disparaging Adams in private letters to members of his own party, and supporting radicals like Cooper, who called Adams a despot, no better than the monarchs of Europe. The purloined Stone letters had strongly implied that Priestley was plotting with America's enemies. Priestley had publicly allied himself with Jefferson and the Republican opposition, despite all his talk about the dangers of political factions. And now Pickering wanted to deport him, or at least put the heat on. The decision lay at a crossroads of great personal and historic magnitude: a man choosing whether to apply the full force of law against his former friend; a young nation wrestling with the question of how it would handle its intellectual dissidents. Would they be tolerated, even protected?

Or would they be silenced? The French had already made it clear where their revolution had taken them: Lavoisier had been executed five years before, during the Reign of Terror. Would the United States embark on the same path with Priestley?

Adams had a reputation for being thin-skinned, but in this one extraordinary instance—faced with undeniable personal betrayal and at least the accusation of public treason—he took another approach. He blinked.

ADAMS WROTE the specific lines that spared Priestley on August 16, in a letter sent back to Pickering: "I do not think it wise to execute the alien law against poor Priestley at present. He is as weak as water, as unstable as Reuben, or the wind. His influence is not an atom in the world." The words were simple enough, but, to borrow a phrase from Whitman, they contained multitudes. Their meaning is so unstable to us now, because they tease us with the answer to two key questions about Adams: his true feelings about Priestley, which indirectly leads us to the more momentous question of his true feelings about the Alien and Sedition Acts. A considerable part of the historical status of the Adams presidency hinges on his relationship to the Acts. Certainly the decision to sign the Acts was crucial to Jefferson's assessment of Adams. Nearly everyone now agrees with Jefferson, that that decision was a mistake; the debate centers on whether it was a mistake that Adams willingly made, or whether it was one that he was forced, against his will, into making—forced by the heightened tensions of political partisanship and the threat of war.

The literal meaning of the lines to Pickering is clear enough: Priestley is an old, confused man, obsessed with the ten horns of the Beast, and geographically isolated from the centers of power, in both America and Europe. He poses no threat to the Union in such a doddering state. But did Adams truly believe this of his old friend? Clearly Priestley's bizarre proclamations over breakfast had made a deep impression on Adams; he would still be writing about Priestley's apocalyptic musings twenty years later. That experience could well have left Adams with the impression that Priestley was unstable, but it still doesn't justify the claim about his limited influence. He was clearly supporting Cooper, who had direct access to the megaphone of the *Aurora*. And Priestley was one of the most distinguished intellectuals in the United States. With Rittenhouse dead, there was no real rival to Priestley in terms of scientific achievement, and as a theologian—at least as measured by the international reach of his work—he had no peers in the new country. Most of all, he had the ear of the vice president. Place all those factors on the scale of influence, and there is no reasonable scenario where they weigh in at "not an atom in this world."

So why would Adams make Priestley sound weaker than he actually was? One potential answer, which Adams would himself suggest in his later correspondence with Jefferson, was that he had signed the Alien and Sedition Acts as a gesture of political conciliation, but had no intention of enforcing them to the full extent of the law. His plan was to sign and then (selectively) undermine. He was after genuine spies, and not the loyal opposition. Joseph Priestley Jr. later reported that Adams had sent a message privately to Priestley after the August 16th note to Pickering, saying that "he wished [Priestley] would abstain from saying anything on politics, lest he should get into difficulty." The president made it clear that Priestley was "one of the persons contemplated when the law was passed," which struck Adams as a sign that the aggressive factions within his party did not understand Priestley's "real character and disposition."

The other possibility is that Adams felt pangs of guilt that centered exclusively on Priestley himself, because of their existing relationship, but was otherwise entirely happy to throw his weight behind the new laws. The disparity between Cooper's treatment and Priestley's makes this the slightly more plausible scenario. Cooper was a polemicist, and far more of a hothead than Priestley, but he was quite clearly not a spy. And yet Adams was entirely willing to send him to jail for six months. If the famous radical theologian in Northumberland didn't pose enough of a threat to justify prosecuting him, why bother incarcerating his deputy?

Whatever his true motivation, Adams spared Priestley the torment of becoming a political prisoner in his adopted homeland. Inspired in part by Adams's suggestion that Priestley's enemies did not understand his "real character," Priestley set out to write a thorough response to his American critics. The final result of that effort, *Letters to the Inhabitants of Northumberland and Its Neighborhood*, would be the last great work of his enormously prolific life. Published at the end of 1799, the *Letters* were divided into two main sections: a long inventory of all the charges against Priestley (his religious unorthodoxies, his support for France, the purloined letters, the Cooper handbills), and then a series of short essays on the political and constitutional questions of the day. The most impassioned section of the *Letters* conveyed Priestley's long-standing support for the colonies' struggle against England, and alluded to his rich friendship and collaboration with Franklin. It was, in a way, a summing up of the past thirty years of his political journey, tracing a line back to the London Coffee House and the Honest Whigs, and their collective dream of a new form of enlightened liberty across the Atlantic. In framing the story of his life this way, Priestley turned the nativist rhetoric of his critics on its head: rather than being traitors or spies, émigrés like Priestley and Cooper, who had come to America voluntarily seeking the promised freedom of the new land, had the *most* investment in seeing the new nation live up to its founding principles:

> To find in America the same maxims of government, and the same proceedings, from which many of us fled from Europe, and to be reproached as disturbers of government there, and chiefly because we did what the court of England will never forgive in favour of liberty here, is, we own, a great disappointment to us, especially as we cannot now return. Had Dr. Price himself, the great friend of American

liberty in England, or Dr. Wren, with both of whom I zealously acted in behalf of your prisoners, who must otherwise have starved, and in every other way in which we could safely serve your cause, because we thought it the cause of liberty and justice, against tyranny and oppression; I say, had either of these zealous, and active, and certainly disinterested, friends of America been now living, they would not have been more welcome here than myself; and they would have held up their hands with astonishment to see many of the old Tories, the avowed enemies of your revolution, in greater favour than themselves.

As had been the case with the Birmingham Letter, the appeal to his Northumberland neighbors did little to quiet the more vitriolic of his critics. But many supporters of the Republican cause considered the *Letters* to be the most stirring and persuasive indictment of the Adams administration on record. Jefferson distributed copies to a dozen of his friends in Virginia, and wrote back to Priestley that the essays, along with Cooper's, had been "the most precious gifts that can be made to us. . . . From the Porcupines of our country you will receive no thanks; but the great mass of our nation will edify & thank you." He made no effort to conceal his visceral disgust with Adams, including a sly reference to the monarchical tendencies many saw in the current president: "How deeply have I been chagrined & mortified at the persecutions which fanaticism & monarchy have excited against you, even here!"

THE TURBULENCE OF the Northumberland years would eventually subside. On March 3, 1801, Thomas Jefferson was sworn in as the third president of the United States. Several weeks later, inspired by news that Priestley had recovered from a serious illness, Jefferson sat down to write a letter to his friend in Northumberland. Rather than distance himself from the eclectic minister, he would embrace him, in what would prove to be one of the most important letters in the immense archive of Jefferson's correspondence.

The letter began with an extraordinary tribute to Priestley himself:

> It was not till yesterday I received information that you . . . had been very ill, but were on the recovery. I sincerely rejoice that you are so. Yours is one of the few lives precious to mankind, and for the continuance of which every thinking man is solicitous.

After these opening salutations, Jefferson quickly shifted into an attack on the abuses of the previous administration and the furor of public opinion than had rained down on Priestley:

> What an effort my dear Sir of bigotry, in politics and religion, have we gone through! The barbarians really flattered themselves they should be able to bring back the

times of Vandalism, when ignorance put everything into the hands of power and priestcraft. All advances in science were proscribed as innovations. They pretended to praise and encourage education, but it was to be the education of our ancestors. We were to look backwards, not forwards, for improvement; the President himself declaring in one of his answers to addresses that we were never to expect to go beyond them in real science. This was the real ground of all the attacks on you.

"All advances in science were proscribed as innovations." Jefferson is using the older, negative sense of the word "innovation" here: a new development that threatened the existing order in a detrimental way. (The change in the valence of the word over the next century is one measure of society's shifting relationship to progress.) But that regressive age was now over, and Priestley—the most forward-thinking mind of his generation—could now consider himself fully at home:

Our countrymen have recovered from the alarm into which art and industry had thrown them; science and honesty are replaced on their high ground, and you, my dear Sir, as their great apostle, are on its pinnacle. It is with heartfelt satisfaction that in the first moments of my public action, I can hail you with welcome to our land, tender to you the homage of its respect and esteem, cover you under the protection of those laws which were made for the wise and good like you, and disdain the legitimacy of that libel on legislation which under the form of a law was for some time placed among them.

Perhaps inspired by the legendary optimism of Priestley himself, Jefferson then added some of the most stirringly hopeful words that he ever put to paper:

As the storm is now subsiding, and the horizon becoming serene, it is pleasant to consider the phenomenon with attention. We can no longer say there is nothing new under the sun. For this whole chapter in the history of man is new. The great extent of our Republic is new. Its sparse habitation is new. The mighty wave of public opinion which has rolled over it is new. But the most pleasing novelty is, its so quietly subsiding over such an extent of surface to its true level again. The order and good sense displayed in this recovery from delusion, and in the momentous crisis which lately arose, really bespeak a strength of character in our nation which augurs well for the duration of our Republic, and I am much better satisfied now of its stability than I was before it was tried.

This is politics seen through the eyes of an Enlightened rationalist. The American experiment was, literally, an experiment, like one of Priestley's elaborate concoctions in the Fair Hill lab: a system of causes and effects, checks and balances, that could only be truly tested by running the experiment with live subjects. The political order was to be celebrated not because it had the force of law, or divine right, or a standing army behind it. Its strength came from its internal balance, or homeostasis, its ability to rein in and subdue efforts to destabilize it.

The inaugural letters made it clear how much each man owed the other: Priestley had shown Jefferson a way out of his religious impasse, providing the intellectual bedrock for Jefferson's Christian faith; he had composed, at great peril to himself, the most rousing defense of Republican values during the Alien and Sedition controversy; in the coming years, Priestley would help Jefferson plan out the curriculum for the new university that would be a key part of Jefferson's intellectual legacy, returning Priestley to his original passion for educational reform. Jefferson, in turn, had been Priestley's great champion inside the Adams administration, and had now offered him, as chief executive, "the protection of those laws which were made for the wise and good."

Greatly moved by Jefferson's letters, Priestley forwarded some of them to Lindsey, with the remark: "[For] the first time in my life (and I shall soon enter my 70th year) I find myself in any degree of favour with the governor of the country in which I have lived, and I hope I shall die in the same pleasing situation." He had lost the companionship of Mary, and the camaraderie of the Honest Whigs and the Lunar Society. But he had, at long last, found a government under which he could live in peace.

PRIESTLEY AND JEFFERSON corresponded regularly during the first few years of the new administration. Priestley published a few scattered scientific papers, some of them still carrying the torch for his phlogiston theory, and he would send them down to Monticello or Washington to ensure that "Politicks not make [Jefferson] forget what is due to Science." Jefferson would urge Priestley to relocate to the milder climate of Virginia: "The choice you made of our country for your asylum was honorable to it; and I lament that for the sake of your happiness and health its most benign climates were not selected." They traded thoughts on how to make the curriculum at the University of Virginia as innovative as possible, "looking forward, not backwards, for improvement," as always.

Even with his health fading, Priestley remained amazingly prolific to the very end, publishing four new volumes in his *General History of the Christian Church* in 1803. But by the beginning of 1804, his chronic battle with indigestion had made him suddenly much more feeble. "Much worse: incapable of business," he wrote in his diary on February 2nd. Three days later, aware that the end was near, he asked each of his grandchildren to visit with him separately at his bedside. "I am going to sleep as well as you," he said to them, "for death is only a good long sound sleep in the grave, and we shall meet again." The next morning he spent dictating corrections to Cooper and his son Joseph, for a batch of new pamphlets they planned to publish. When they read back the changes, he nodded in assent: "That is right; I have now done." Forty minutes later, he was dead.

A few days before his death, Priestley had sent a letter to

a friend with one last request. "Tell Mr. Jefferson," he wrote, "that I think myself happy to have lived so long under his excellent administration; and that I have a prospect of dying in it. It is, I am confident, the best on the face of the earth, and yet I hope to rise to some thing more excellent."

PRIESTLEY WAS BURIED in a Quaker cemetery in Northumberland. Many eulogies and tributes followed the news of his demise, as it slowly spread around the world. The American Philosophical Society held a memorial service in Philadelphia. The parishioners of the New Meeting House in Birmingham, built over the ruins of the church destroyed in the riots more than a decade before, wore mourning clothes for two months. From the pulpit at Mill Hill Chapel in Leeds, Priestley's successor called him a "burning and shining light."

Yet Priestley's maverick beliefs and cross-disciplinary thinking would damage his reputation in the coming decades. He became a kind of sacrificial lamb for the parallel developments of specialization and professionalization that dominated nineteenth-century science. Serious science became the province of experts and specialists, not dabblers and amateurs. Pioneering research—according to the new consensus—required that the scientist isolate himself from the external worlds of politics or faith, and not seek connections to them. The first volley of that attack arrived a few days before Priestley's death, in a brief, caustic item that appeared in the *Times* of London: "Dr. Priestley's health is said to be in a declining state; his reputation has long been. He left England for America, in search of Liberty, and has been laughed at by the Americans for his folly. Such is the natural, and merited close of a man's life, who, a Christian teacher and a philosopher, left the highways of religion and science, for the crooked paths of politics." But the more substantive rendition arrived a few years later in the *Edinburgh Review*, part of a lengthy assessment of Priestley's memoirs, which had been published, along with some hagiographic commentary, by Thomas Cooper in 1805. The review scolded Priestley for his adventures into politics, but it also mounted a direct assault on his scientific method:

He had great merit in the contrivance of his apparatus, which was simple and neat, to a degree that has never been equalled. . . . The truth is, however, that he was always too much occupied with making experiments, to have leisure either to plan them beforehand with philosophical precision, or to combine their results afterwards into systematic conclusions. . . . [He] seems to have been entirely forgetful of Bacon's invaluable precepts, that experiments should not be many, but decisive, and that they should be preceded by certain limited hypotheses or conjectures. . . . Without these precautions . . . to make experiments, however numerous or however pretty, was merely to grope in the dark, and could scarely ever lead to valuable or certain conclusions. The greater part of

Dr. Priestley's experiments are exactly of this description. There is about as much philosophy in them, as in sweeping the sky for comets.

The great French naturalist Georges Cuvier penned a more generous eulogy that nonetheless pointed to the same failings, along with Priestley's stubborn refusal to abandon phlogiston: "He was the father of modern chemistry," Cuvier famously wrote, "who never acknowledged his daughter." Many formal accounts of Enlightenment science composed in the nineteenth century struggled to make sense of Priestley's eccentric career. The entry on Priestley in the *Dictionary of National Biography* described his research as "often superficial."

Over time, though, the tide of opinion began to turn, led in part by the rise of environmental science in the second half of the twentieth century. In 1922, the American Chemical Society established the Priestley Medal for "distinguished service in the field of chemistry." Statues and plaques in Leeds, Birmingham, and Northumberland now mark the important milestones in his life. (Though no memorial records the location of the London Coffee House, the site of so much Enlightenment-era inspiration.) The lab at Bowood house where he isolated oxygen became one of the first chemical landmarks named by the American Chemical Society in the early 1990s.

More important, though, the values that Priestley brought to his intellectual explorations have never been more essential

than they are today. The necessity of open information networks—like ones he cultivated with the Honest Whigs and
the Lunar Society, and with the popular tone of his scientific
publications—has become a defining creed of the Internet
age. That is in part because the flow of information dif-
fers from the flow of energy in one crucial respect: there

is a finite supply of energy, which means that tapping it is
invariably a zero-sum game. (Burning Carboniferous fuel in
steam engines during the eighteenth century leaves less in
the ground for the twenty-first.) But the spread of information does not come with the same cost, particularly in the
age of global networks. An idea that flows through a society
does not grow less useful as it circulates; most of the time,
the opposite occurs: the idea improves, as its circulation
attracts the "attention of the Ingenious," as Franklin put
it. Jefferson saw the same phenomenon, and interpreted it
as yet another part of nature's rational system: "That ideas
should freely spread from one to another over the globe,"
he wrote in an 1813 letter discussing a patent dispute, "for
the moral and mutual instruction of man, and improvement
of his condition, seems to have been peculiarly and benevolently designed by nature, when she made them, like fire,
expansible over all space, without lessening their density at
any point, and like the air in which we breathe, move, and
have our physical being, incapable of confinement or exclusive appropriation."

That new openness has helped nurture the kind of multidisciplinary thinking that was the hallmark of Priestley's

intellect. Fields like information theory, ecosystem science, and evolutionary theory rank among the most influential and generative scientific fields of the past fifty years, spawning debates that have unavoidable consequences for the spheres of politics and of faith (even if the presidential candidates usually try to avoid them). Priestley would have grasped immediately how so many of today's discoveries are bound up in social and political affairs: global warming, stem-cell research, intelligent design, neuroscience, atomic energy, the genomic revolution, not to mention the massive social disruptions introduced by computer science in the form of the Internet. Building a coherent theory of the modern world *without* a thorough understanding of that science would have struck Priestley as a scandal of the first order.

To be sure, the rising peaks of scientific progress means that specialization is an unavoidable reality: the facts are so much more complex than they were in Priestley's day, thanks to two centuries of empirical research. Amateur chemists are not likely to discover new elements in their home laboratories anymore, which is itself a sign of progress. And Priestley's critics from the nineteenth century had a legitimate point about the limits of his method: it took Lavoisier's more systematic approach to define the new paradigm of modern chemistry. But to stop there is to miss the distinct kind of contribution that Priestley brought to the many fields he explored. That roving, untutored, connective intelligence was not particularly suited for defining the bylaws of a new scientific paradigm. But it was exceptionally well suited for *exploding* the old conventions, for pushing the field into its revolutionary mode. Some great minds become great by turning the rubble of an exploded paradigm into something consistent and meaningful. Others become great by laying the gunpowder, grain by grain. Every important revolution needs both kinds of minds to complete itself. Priestley himself grasped this quality in his work more clearly than either his critics or his disciples: "It may be my fate to be a kind of comet, or flaming meteor in science," he wrote in 1775, "in the regions of which (like enough to a meteor) I made my appearance very lately, and very unexpectedly; and therefore, like a meteor, it may be my destiny to move very swiftly, burn away with great heat and violence, and become as suddenly extinct."

NEARLY A DECADE after his death, that comet would sweep across the American sky one last time, and in doing so transform one of the great conversations in the history of political thought: the Jefferson-Adams letters. The falling out between the two ex-presidents had been so severe that the men lost all contact with each other for a decade, save one fitful and tense exchange in 1804, when Jefferson briefly corresponded with Abigail Adams. In early 1812, however, at the urging of their mutual friend Benjamin Rush, Adams and Jefferson began corresponding again, with Adams sending off the first amiable letter. It was the beginning of a conversation that would last another fourteen years, two aging patriarchs debating the meaning and future prospects of the grand American experiment that they themselves had engineered. It lasted all the way to that most implausible of endings: both men dying on July 4, 1826, the fiftieth anniversary of the signing of the Declaration of Independence.

The first year of their correspondence, however, lacked the passion and engagement and argument that would ultimately make it so fascinating. There is a sense of careful decorum and fragility to these first exchanges, as if the two men were tiptoeing through a minefield of their past hostilities. Much of the prose remains focused on the personal domain: they pass on inventories of their grandchildren and great-grandchildren; complain about their failing health; tally up the number of Declaration signatories still alive. There is some talk about contemporary politics, and a few fond references to their collaborations in the 1770s, but almost no allusions to the turbulence and rancor that would follow.

All that would change, though, with the publication of a book in London in 1812: the posthumous memoirs of the Reverend Theophilus Lindsey, including a generous appendix of "Letters of eminent Persons, his Friends and Correspondents." In that collection were Jefferson's post-inaugural letters to Priestley, which Priestley had forwarded to Lindsey in confidence a decade before. Somehow a copy found its way to Adams in Quincy in May of 1813. Adams read through the letters, and all the old anger and resentment from that period boiled over in him again. He wrote a quick note to Jefferson, asking if he was familiar with the volume, and promising that he would have more to say. Ten days later, he was back, this time quoting Jefferson's letter in detail: "We were to look backwards, not forwards, for improvement; the President himself declaring in one of his answers to addresses, that We were never to expect to go beyond them in real Science." Adams vehemently denied ever uttering such a statement: "The sentiment you have attributed to me in your letter to Dr. Priestley I totally disclaim and demand in the French sense of the word demand of you the proof. It is totally incongruous to every principle of my mind and every Sentiment of my heart for Threescore Years at least."

Four days later, on June 14, he fired off another screed, this time quoting Jefferson's reference to the Alien and Sedition Acts as a "Libel on legislation."

As your name is subscribed to that law as Vice President, and mine as President, I know not why you are not as responsible for it as I am. Neither of us was concerned in the formation of it. We were then at war with France. French spies then swarmed in our cities and our country; some of them were intolerably impudent, turbulent, and seditious. To check these was the design of this law. Was there ever a government which had not authority to defend itself against spies in its own bosom, spies of an enemy at war. This law was never executed by me, in any Instance.

The next day, Jefferson wrote his first reply, a long and gracious letter, attempting to soothe his combustible friend: "[The letter] recalls to our recollection the gloomy transactions of the times, the doctrines they witness, and the sensibilities they excited. It was a confidential communication of reflections of these from one friend to another, deposited in his bosom, and never meant to trouble the public mind." Yet Jefferson would not concede everything. "Whether the character of the times is justly portrayed, posterity will decide. But on one feature of them they can never decide, the sensations excited in free yet firm minds, by the terrorism of the day. None can conceive who did not witness them, and they were felt by one party only."

With those lines, the exchange between Adams and Jefferson became a genuine, two-way debate. "It was," the historian Joseph Ellis writes, "the defining moment in the correspondence," the point at which it "became an argument between competing versions of the revolutionary legacy." It would rage in its most heated form for the next four months, driven by a constant barrage of more than twenty agitated letters from Quincy, interrupted by five longer and more contemplative replies from Monticello. They discuss the inevitability of political parties, and the "terrorism" of the Alien and Sedition period. Adams burrows through his own personal archive to track down the speech where he had denounced the innovations of science, arguing that Jefferson had misunderstood the original context. Adams offers a quote from his own "Defense of the Constitutions" (published in 1787) that lauds the "Invention of Mechanic Arts" and the "discoveries in Natural Philosophy." The two men ponder why systems of government have not progressed at the same speed as natural philosophy. And they delve deeply into Priestley's unorthodox vision of Christianity and its influence on Jefferson. This is no ordinary conversation, not just because it involves the two great living patriarchs of the American Revolution, but also because it transpires through an extremely unusual, almost postmodern, literary device: nearly all of Adams's letters pivot off of specific quotes from Jefferson's original exchange with Priestley. At the heart of this great American conversation, then, we find a strange sort of deconstruction taking place, with Adams meticulously unpacking, sentence by sentence, the turns of phrase that Jefferson had written more than a decade earlier. The whole context seems, to the modern reader, like something from a Borges short story or a Calvino novel, or one of the layered epistolary novels of the eighteenth century: a letter is written, then forwarded, then published, then discovered by one of the people vilified in the original text, then forwarded again back to its original author, with extensive annotations. That palimpsest of commentary upon commentary is what ignited the most epic conversation in American history. And there, at the center of that textual web, almost ten years after his death, lay Joseph Priestley.

THE FACT THAT PRIESTLEY should play such a transformative role in the Jefferson-Adams letters, coupled with the fact that he is mentioned in that archive far more frequently than Washington, Franklin, or Madison, gives us some sense of the magnitude of Priestley's presence in the minds of Jefferson and Adams. Priestley was a kind of Zelig of early American history, appearing at key turning points like some kind of errant founding father: Franklin's kite; the Privy Council; Alien and Sedition; the Jefferson-Adams correspondence. One of the final letters Adams wrote, at the age of eighty-eight, recounted in exacting detail the breakfast he had with Priestley almost thirty years before, where Priestley had waxed apocalyptic in his interpretation of the French Revolution. Decades after his death, Adams and Jefferson were still debating the ideas that their old friend had unleashed on the world. "This great, excellent, and extraordinary Man, whom I sincerely loved, esteemed, and respected, was really a Phenomenon: a Comet in the System, like Voltaire," Adams wrote. Priestley's ideas lived on so vividly in the Jefferson-Adams correspondence because they were, in multiple ways, central to the American experiment itself.

What, then, does Priestley's life tell us about the great paradigm shift of the American Revolution? We have been debating what the founders stood for practically since the ink dried on Jefferson's first draft of the Declaration. But something different happens when you look at the birth of America through the outsider view of Priestley's career—when you take Jefferson at his word that Priestley's life was "one of the few precious to mankind," when you think of Franklin's longing to return to his happiest days, trading ideas at the London Coffee House. If Priestley was so central a figure to the three towering intellects central to the birth of the United

States, how does that shape our perception of the founders—and the values they pass on to us today?

Clearly one lesson is that Priestley—and his kindred spirits in London, Birmingham, Quincy, and Monticello—refused to compartmentalize science, faith, and politics. They saw those three systems not as separate intellectual fiefdoms, but rather as a continuum, or a connected web. The new explanations of natural philosophy could help shape new political systems and redefine faith for an Enlightened age. Adopting a know-nothing attitude toward scientific understanding—to hide behind the cloak of piety or political dogma—would have been the gravest offense to Priestley and his disciples. It is no accident that, despite the long litany of injuries Adams felt had been dealt him in Jefferson's letters to Priestley, he chose to begin his counterassault by denying, as a point of honor, that he had ever publicly taken a position as president that was resistant to the innovations of science. Remember that Jefferson had also insinuated that Adams had betrayed the Constitution with his "libel on legislation." But Adams lashed out first at the accusation that he was anti-science. That alone tells us something about the gap that separates the current political climate from that of the founders.

In the popular folklore of American history, there is a sense in which the founders' various achievements in natural philosophy—Franklin's electrical experiments, Jefferson's botany—serve as a kind of sanctified extracurricular activity. They were statesmen and political visionaries who just happened to be hobbyists in science, albeit amazingly successful ones. Their great passions were liberty and freedom and democracy; the experiments were a side project. But the Priestley view suggests that the story has it backward. Yes, they were hobbyists and amateurs at natural philosophy, but so were all the great minds of Enlightenment-era science. What they shared was a fundamental belief that the world could change—that it could *improve*—if the light of reason was allowed to shine upon it. And that belief emanated from the great ascent of science over the past century, the upward trajectory that Priestley had so powerfully conveyed in his *History and Present State of Electricity*. The political possibilities for change were modeled after the change they had all experienced through the advancements in natural philosophy. With Priestley, they grasped the *political* power of the air pump and the electrical machine.

We like to talk about the American sensibility in terms of its inveterate optimism, but when one reads Jefferson, Franklin, and Adams, what one finds in each man is a slightly different streak of darkness: Franklin was a borderline misanthrope; Jefferson was horrified by the emerging power of cities and industrialization; Adams had his furies and his nagging sense that the world was not respecting his achievements. The temperament that we expect to find at the birth of America—bountiful optimism, an untroubled sense that the world must inevitably see the light of reason—arrives aboard the *Samson* in 1794. Priestley seems to have had a remarkable capacity to bring out the most positive feelings in his friends, as in Jefferson's post-inaugural letter ("This whole chapter in the history of man is new"). He was a true progressive, in the literal meaning of the term, in that he thought the world was headed naturally toward an increase in liberty and understanding, what he called the "sublime" view in the introduction to *The History*:

> For an object in which we see a perpetual progress and improvement is, as it were, continually rising in its magnitude; and moreover, when we see an actual increase, in a long period of time past, we cannot help forming an idea of an unlimited increase in futurity; which is a prospect really boundless, and sublime.

That faith in progress was challenged by riots, exile, and the threat of prosecution, but it survived to his last days, under the "excellent administration" of Thomas Jefferson.

The faith in science and progress necessitated one other core value that Priestley shared with Jefferson and Franklin, and that is the radical's belief that progress inevitably undermines the institutions and belief systems of the past. (Whether Adams truly shared this perspective is a more complicated question, one that was central to the initial flare-up in the correspondence with Jefferson.) To embrace the sublime vista of reason was, inevitably, to shake off a thousand old conventions and pieties. It forced you to rewrite the Bible, and contest the divinity of Jesus Christ; it forced you to throw out all the august, Latinate traditions of the educational establishment; it forced you to invent whole new modes of government; it forced you to think of the air we breathe as part of a natural system that could be disturbed by human intervention; it forced you to dream up entirely new structures for the transmission and cultivation of ideas. You could no longer put stock in "the education of our ancestors," as Jefferson derisively called it. Embracing change meant embracing the possibility that everything would have to be reinvented.

All of these values exist separately today on various points of the political spectrum. But to find them strung together as a single, unified worldview is astonishingly rare. We have always had a steady supply of politicians who speak euphorically about the great possibilities that lie ahead, and just as many who connect that sense of hope to their religious values. But, ironically, the vision of "morning in America" usually involves a return to simpler times, the old conventions, the education of our ancestors. Those who still argue for the possibility of radical change—in government, in faith, in our economic systems—increasingly center their arguments on the bedrock of scientific understanding, largely the ecosystem science that Priestley helped invent. But the radical's default temperament today is precisely the opposite of Priestley's: bleak and dystopian, filled with gloomy predictions of imminent catastrophe. To be a progressive today is to believe that the great engine of progress has stalled, and that we are no longer climbing the mountain, but descending into a valley of self-destruction.

It is possible that the circumstances of our age do, in fact, warrant these views. Perhaps the Priestley worldview is obsolete for a reason. Perhaps the era of radical change has passed us by, or the steady march of progress has reversed itself. Yet one thing is clear: to see the world in this way—to disconnect the timeless insights of science and faith from the transitory world of politics; to give up the sublime view of progress; to rely on the old institutions and not conjure up new ones—is to betray the core and connected values that Priestley shared with the American founders. Thanks to the accelerating march of human understanding, we now see the web of relationships far more clearly than Priestley or Franklin or Jefferson could: we can link a single molecule of oxygen; the biochemical engine of photosynthesis; the atmospheric explosion of breathable air; the immense energy deposits of the Carboniferous era; the rise of industrialization; the political turmoil of Priestley's day; and the environmental crisis of our own. All those elements now exist for us as a connected system, understood with a level of precision and subtlety that would have delighted Priestley, though not surprised him, given his expectations. How can such a dramatically expanded vista not make us think that the world is still ripe for radical change, for new ways of sharing ideas or organizing human life? And how could it not also be cause for hope?

# ACKNOWLEDGMENTS

I'm grateful to several institutions for their willingness to let me work through the major themes of this book in public. First, NYU's School of Journalism, for letting me teach a graduate seminar on Cultural Ecosystems, and my students there who contributed so many helpful ideas (and who, I'm thankful to report, shot down more than a few of my less helpful ones). My friends at the Long Now Foundation—Stewart Brand, Kevin Kelly, Brian Eno, Danny Hillis, and Alexander Rose—were kind enough to invite me to discuss the "long zoom" approach to cultural history at one of their seminars in long-term thinking in 2007. I was also lucky enough to be invited to discuss these issues onstage with Brian at the Institute of Contemporary Arts in London. I'm also indebted to Larry Lessig for the Jefferson quote at the beginning of this book, an early link that led me to one of the book's major themes. But the most important intellectual debt I owe for this book is to my cousin Jay Haynes, with whom I have been discussing new political models, environmentalism, and other heresies for almost thirty years now. So thanks to all these folks: you are truly my heroes, my own merry band of Honest Whigs. Go on and prosper.

Speaking of the Honest Whigs, I can't fail to mention the coffeehouses that made this book possible: first the late lamented Tea Lounge in the South Slope, then Brooklyn's unparalleled Gorilla Coffee. (Even better than the coffee, though, were the many walks to and fro with Mark Bailey.) My colleagues and investors at outside.in were incredibly supportive of my admittedly unorthodox desire to write a book while running an Internet startup; I'm particularly indebted to Mark Josephson, Rob Deeming, John Geraci, Cory Forsyth, Andy Karsch, Fred Wilson, Ed Goodman, George Crowley, Richard Smith, and John Borthwick for their support.

My research assistant, Jared Ranere, responded with aplomb to the most bizarre of my requests—"Giant dragonflies!"—and was an invaluable sounding board from start to finish. I'm grateful for the institutions that supported the research for this book: the University of Chicago Library; the New York Public Library; the NYU Library; the Jefferson Library at Monticello; the Papers of Benjamin Franklin, maintained by the American Philosophical Society and Yale; the Avalon Project at Yale Law Library; the Royal Society; the British Library; the Priestley collection at the University of Pennsylvania. I also greatly benefited from the many public-domain works by Priestley and his peers that are now available as full-text documents from Google Books.

Thanks to the many people who read this manuscript in draft: Oliver Morton, Garry Wills, Kurt Andersen, Walter Isaacson, David Smith, Malcolm Dick, and Alexa Robinson. Any errors that remain are, of course, entirely of my own making. Special thanks to the good people at Riverhead, led by my longtime editor, Sean McDonald, who handled the
sudden emergence of this book idea with grace and keen understanding, and who supported me through the chaotic final weeks of putting it together. Thanks, too, to Geoff Kloske, Matthew Venzon, Hal Fessenden, Meredith Phebus, Marie Finamore, and Emily Bell. Once again, my agent, Lydia Wills, helped steer the ship of my career with her amazing sense of direction, despite a few unusual detours this time around.

This was the most peripatetic of all my books. Parts of it were written in Brooklyn; on Martha's Vineyard; on the coast of northern California; in London, Brighton, and Bath in the UK. But I wrote most of it on Shelter Island, looking out through the two forks of Long Island toward the Atlantic Ocean, following the Gulph Stream's righthand turn toward England. Thanks to all the friends and family who accompanied me on those journeys or who made them possible, in particular: Donald and Mary Fraser, for their writing nook on Shelter; Paul Miller, Birmingham tour guide par excellence; and David Smith, who spent a lovely morning with me at Bowood House.

Finally, there's my family: Alexa Robinson and our boys, Clay, Rowan, and Dean. Joseph Priestley lived in a world dramatically different from the one I live in, but the one aspect of his life that seems immediately familiar lies in his descriptions of life at home with Mary and the kids: writing in a house filled with the boisterous play of children; the daily intellectual camaraderie of sharing ideas with a lifelong partner. (Not to mention all the games!) Thanks to the four of you, as always, for your love and inspiration.

September 2008
Brooklyn

# NOTES

## Author's Note

xvii *president of the United States* The candidate in question was the ordained minister Mike Huckabee, who finished second in the race for the Republican nomination. Huckabee was speaking at an early presidential debate in June of 2007. He went on to add: "If they want a president that doesn't believe in God, there's plenty of choice. My point [on the question of evolution] is, I don't know. I wasn't there. But I believe whether God did it in six days or whether he did it in six days that represented periods of time, he did it. And that's what's important. But you know, if anybody wants to believe they're the descendants of a primate, they're welcome to do it."

## Prologue: The Vortex

4 *"a kind of column"* Quoted in Lundy, p. 203.

6 *"But during the voyage"* Mittelberger, p. 24.

7 *"We had many things to amuse us"* Quoted in Jackson, p. 310.

8 *"Our voyage at times was very unpleasant"* Quoted in Moser, p. 21.

9 *"Inquisitions and Despotisms are not alone"* John Adams to Joseph Priestley, February 19, 1792. Quoted in Graham, p. 177.

11 *"the waters mov'd away"* De Vorsey, p. 106.

## Chapter One: The Electricians

18 *"It consists of clergymen"* Griffith, p. 5.

18 *His first trip to London* Priestley 1904, p. 19.

20 *"This proved a very suitable and happy connexion"* Ibid., p. 30.

22 *the many years Franklin spent poring over balance sheets* Isaacson, p. 137.

23 *"There is something however in the experiments of points"* Benjamin Franklin to Peter Collinson, March 2, 1750.

24 *"A man in Philadelphia in America"* Van Doren, p. 170.

25 *"You will find [Priestley] a benevolent, sensible man"* Quoted in Schofield 1966, p. 14.

26 *"Much was said this night"* Crane, p. 229.

26 *"a little of his preparation"* Ibid., p. 224.

28 *"No other work known to the history of science"* Kuhn, p. 30.

28 *"presented discovery as a set"* Shaffer 1986, p. 207.

31 *"I have made an experiment"* Quoted in Schofield 1966, p. 15.

32 *"I took a cork"* Ibid., p. 21.

32 *"I have made a great number of experiments"* Ibid., p. 35.

36 *"Were it possible to trace"* Priestley 1775, p. xv.

37 *"The History of Electricity"* Ibid., p. i.

38 *"to look down from the eminence"* Ibid., p. iv.

39 *"To demonstrate, in the completest manner possible"* Ibid., p. 160.

43 *"Nothing ever happened in baseball"* Gould, p. 466.

46 *contention that class identity, capital, and technological accelera-tion would be prime movers in the coming centuries . . . in ways that the original inventors never anticipated* Marx just failed to predict correctly where they were all taking us as a society, in that he thought the dialectical progression of history was leading to the ultimate synthesis of a true communist state. In part his prediction failed because he neglected other macro forces, including the capacity of capitalism to evolve corrections to the problems it created, and in part because he couldn't shake off the organizing principle of Hegel's dialectic.

48 *"Aside from occasional brief asides"* Kuhn, p. xii.

51 *thousands (or millions) of years to play out* This layered view of cultural development was directly inspired by the pace lay-ered diagram of civilization that I first encountered in Stew-art Brand's wonderful book, *How Buildings Learn*. Brand's levels are slightly different, and are focused primarily on the speed at which each layer changes. The main categories are, going from fast to slow: Fashion; Commerce; Infrastructure; Governance; Culture; Nature.

54 *"By the way"* Schofield 1966, p. 54.

59 *"The impact of the introduction of coffee"* Standage, p. 135.

60 *"In electricity, in particular"* Priestley 1775, p. xii.

68  *"The work of a button"* Journal of Jonathan Williams, Jr., of His Tour with Franklin and Others through Northern England, May 28, 1771. Franklin, *The Papers of Benjamin Franklin.*

68  *"made some very pretty Electrical Experiments"* Ibid.

69  *"When I want to admit a particular kind of air"* Priestley 1790, p. 34.

71  *"The plant was not affected"* Priestley 1790, vol. 3, p. 250.

73  *"I have just received the enclos'd"* Benjamin Franklin to John Canton, August 15, 1771.

74  *"[he] had very little knowledge of air"* Priestley 1790, p. xx.

74  *what was there to investigate?* See Shapin and Schaffer's superb *Leviathan and the Air-Pump* for more on the way the air pump transformed the science of pneumatic chemistry and helped define the now conventional notion of scientific experiments.

80  *David Hartley, whose model of cognitive "vibrations" anticipated the modern theory of neuronal association* Schofield summarizes the Hartley/Priestley model of vibrations: "If two or more different vibrations occur at the same time, they will modify each other such that if any one takes place, another, or others, will be excited also, until finally one has a set of fixed vibrational tendencies that respond as triggered by any one of the set . . . Hence simple ideas, by association, become complex, and these more complex still, to produce, with experience and over time, all of our ideas, pleasures, and passions." Schofield, 2004, p. 55.

82  *"as soon as possible"* Rutt 2003, p. 344.

84 *"it burned perfectly"* Priestley 1790, vol. 3, p. 250.

84 *"Several times I divided the quantity of air"* Ibid.

85 *"You may depend on the account I sent you"* Schofield 1966, p. 86.

86 *"I presume that by this time"* Joseph Priestley to Benjamin Franklin, July 1, 1772.

88 *the cocktail of sunlight and oxygen was deadly* Margulis and Sagan, p. 108. "This was by far the greatest pollution crisis the earth has ever endured. Many kinds of microbes were immediately wiped out. Oxygen and light together are lethal—far more dangerous than either by itself. They are still instant killers of those anaerobes that survive in the airless nooks of the present world."

89 *"That the vegetable creation should restore the air"* Benjamin Franklin to Joseph Priestley, July 1772. We do not know the exact date of this letter, because no record of it exists beyond this excerpt from it that Priestley published.

91 *"I hope this will give some check to the rage"* Ibid.

92 *"Once any quantity of air has been rendered noxious"* Priestley 1790, vol. 3, pp. 255–56.

94 *"I present you with this medal"* Rutt 2003, p. 194.

95 *"My Way is"* Benjamin Franklin to Joseph Priestley, Sept. 19, 1772.

96 *"Lord Shelburne is a statesman"* Quoted in Jackson, p. 122.

98 *"I never make the least secret"* Priestley 1904, p. 109.

99 *"Though it was taken out seemingly dead"* Ibid.

99 *"The feeling of it to my lungs"* Ibid.

101 *"More is owing to what we call chance"* Ibid., pp. 102–3.

104 *"It can be taken as an axiom"* Jackson, p. 187.

104 *"Burning added weight"* Ibid.

105 *"Ignoring Scheele"* Kuhn, p. 55

109 *albeit one with an asterisk* The irony is that Priestley introduced his great discovery with a discourse on blind spots. He
failed to recognize that the crucial error lay at the end of his
reasoning, not the beginning.

112 *"As he read the addresses"* Rutt 2003, p. 210.

113 *"I am sorry that the political world"* Priestley 1790, vol. 1,
p. xxvii.


## Chapter Three: Intermezzo

117 *older than the first dinosaurs* For a more extensive account of
*Meganeura* and the oxygen explosion of the Carboniferous,
see David Beerling's superb book *The Emerald Planet.*

120 *"early after life began"* Lovelock and Margulis, p. 2.

123 *"global indigestion"* Beerling, p. 50.


## Chapter Four: The Wild Gas

135 *"You will have heard"* Benjamin Franklin to Joseph Priestley,
May 16, 1775.

136 *"At present am extremely hurried"* Ibid. The modern reader
may be entertained to see that the BlackBerry style of drop-
ping subject pronouns for speed ("At present am extremely
hurried") was alive and well in an age where messages across
the Atlantic took three months to reach their recipient.

137 *"In one of your letters"* Joseph Priestley to Benjamin Franklin, February 13, 1776.

138 *"Though you are so much engaged"* Joseph Priestley to Benjamin Franklin, September 27, 1779.

138 *"I should rejoice much"* Benjamin Franklin to Joseph Priestley, June 7, 1782.

141 *"John Hyacinth Magellan"* Jackson, p. 132. Magellan's real name was Joao Jacinto de Magalhaes, though he adopted the Anglicized version during his London travels.

142 *"Our want of powder"* Kelly, p. 158.

143 *"By Yorktown"* Jackson, p. 202.

143 *"It can truthfully be said"* Kelly, p. 165.

144 *"to the imminent hazard of our most valuable commerce"* Quoted in Schofield 2004, p. 17.

144 *"Ah, Priestley. An evil man, Sir."* Quoted in Kramnick, p. 4.

144 *"Our zeal"* Joseph Priestley to Benjamin Franklin, February 13, 1776.

146 *"If Doctor Priestley applies to my librarian"* Quoted in Schofield 2004, p. 21.

147 *"[The] only method of attaining to a truly valuable agreement"* Ibid., p. 27.

148 *"This rapid process of knowledge"* Priestley 1790, p. xxiii.

152 *"to do more business"* Schofield 1966, p. 204.

154 *"They tried to dine at two o'clock"* Uglow, p. 124.

154 *"Our good friend, Dr. Darwin"* Quoted in Wedgwood, p. 277.

155 *"I am as rich as I wish to be."* Quoted in Gibbs, p. 139.

156 *"I had indeed Thoughts"* Benjamin Franklin to Richard Price, August 16, 1786.

157 *"I know of no Philosopher"* Benjamin Franklin to Joseph Priestley, July 29, 1786.

157 *"Remember me affectionately"* Franklin to Benjamin Vaughan, October 24, 1788.

158 *"I spent the Day"* The Diary of John Adams, April 19, 1786. Interestingly, while Priestley appears to have not met Thomas Jefferson for another ten years, their paths almost crossed in London that April. After recording his first encounter with Priestley, Adams spent the following day with Jefferson, visiting an estate outside of London. Adams, *Adams Family Papers.*

166 *political power . . . south of the imaginary line between Bristol and London* See http://www.ancestry.com/learn/library/article.aspx?=8707.

170 *"The idolatry of the Christian church"* Priestley 1871, p. 103.

171 *"If I have succeeded in this investigation"* Ibid., p. xi.

173 *"To me, [Franklin] acknowledged"* Rutt 2003, p. 212.

174 *"I have read [Priestley's] Corruptions of Christianity"* Thomas Jefferson to John Adams, August 22, 1813.

174 *"I am a Christian"* Thomas Jefferson to John Adams, August 22, 1813.

176 *"a long-lost time and place"* Ellis, pp. 36–37.

176 *"extraordinary attempt . . . to unsettle the faith"* Quoted in Gibbs, p. 172.

177 *"Unitarian principles are gaining ground"* Joseph Priestley and Richard Price, pp. 101–102.

179 *"trade of the good town"* Gibbs, p. 174.

179 *"the liberty, both of that country and America"* Garrett, p. 57.

179 *"The wild* gas*"* Burke, p. 8.

181 *"Don't you remember what a parlous"* Quoted in Gibbs, p. 184.

182 *"Whatever the* modern republicans *may imagine"* Quoted in Gibbs, p. 198.

183 *"A local artist"* Thor, p. 127.

185 *"where they said they meant to broil"* Gibbs, p. 207.

187 *"Accordingly we set off"* Quoted in Thorpe 1906, pp. 131–33.

189 *"Undaunted he heard the blows"* Russell's account is quoted at length in Thorpe, pp. 127–33.

189 *"After living with you 11 years"* Priestley 1791, p. 130.

190 *"You have destroyed"* Priestley 1791, p. 3.

191 *"I shall be obliged to you"* Schofield 1966, p. 262.

192 *"We . . . vow to restore"* Schofield 1966, p. 257.

## Chapter Five: A Comet in the System

198 *"The name of Joseph Priestley will be long remembered"* Graham, p. 49.

199 *"If I had but Fortunatus's wishing-cap"* Ibid., p. 63.

199 *"Whether it be the effect of general liberty"* Ibid., p. 67.

201 *"I am happy and thankful"* Moser, p. 17.

201 *"I could now give a great deal"* Ibid., p. 61.

202 *"Could we have a Coach"* Ibid., p. 76.

203 *"learned, ingenious, and useful"* John Adams to Abigail Adams, March 13, 1796.

203 *"The [Discourses] will be printed"* Ibid.

204 *Priestley and Adams had breakfast alone together* There is much

interesting speculation about the exact date of this breakfast. See Graham, footnote on p. 95.

204 *"My opinion is founded altogether upon revelation and the prophecies"* John Adams to Thomas Jefferson, August 15, 1823.

205 *"She had taken much pleasure"* Quoted in Gibbs, p. 231.

206 *"Having always been very domestic"* Mosner, p. 12.

206 *"a mere hut"* Graham, p. 95.

212 *"I may be doing wrong in writing so freely"* Ibid., p. 117.

213 *"I cannot help thinking that of late years"* Ibid., p. 122.

214 *"A meaner, a more artful, or a more malicious libel"* Ibid., p. 123.

215 *"I do not think it wise"* John Adams to Thomas Pickering, August 16, 1798.

218 *"To find in America the same maxims of government"* Priestley 1826a, p. 167.

219 *"the most precious gifts"* Thomas Jefferson to Joseph Priestley, January 18, 1800.

220 *"It was not till yesterday"* Thomas Jefferson to Joseph Priestley, March 3, 1801.

223 *"[For] the first time in my life"* Quoted in Graham, p. 184.

224 *"The choice you made of our country"* Thomas Jefferson to Joseph Priestley, June 19, 1802.

225 *"Tell Mr. Jefferson"* Quoted in Graham, p. 164.

226 *"Dr. Priestley's health"* London Times, January 28, 1804, p. 2.

226 *"He had great merit in the contrivance of his apparatus"* Monthly Review, pp. 150–51.

227 *"He was the father of modern chemistry"* Cuvier, pp. 209–31.

228 *"That ideas should freely spread"* Thomas Jefferson to Isaac McPherson, August 13, 1813.

230 *"It may be my fate to be a kind of comet"* Quoted in Gibbs, p. 96.

232 *"The sentiment you have attributed to me"* John Adams to Thomas Jefferson, June 10, 1813.

232 *"As your name is subscribed to that law"* John Adams to Thomas Jefferson, June 14, 1813.

233 *"Whether the character of the times is justly portrayed"* Thomas Jefferson to John Adams, June 15, 1813.

233 *"the defining moment in the correspondence"* Ellis 2002, p. 230.

235 *"This great, excellent, and extraordinary Man"* John Adams to Thomas Jefferson, July 18, 1813.

238 *"For an object in which we see a perpetual progress"* Priestley 1775, pp. i–ii.

# BIBLIOGRAPHY

Adams, John, et al. *Adams Family Papers*. Electronic archive maintained by the Massachusetts Historical Society. http://www.masshist.org/digitaladams/aea/.

Badash, Lawrence. "Joseph Priestley's Apparatus for Pneumatic Chemistry." *Journal of the History of Medicine and Allied Sciences* 19 (1964): 139–55.

Balling, R. C., Jr., R. S. Vose, and G. Weber. "Analysis of Long-term European Temperature Records: 1751–1995." *Climate Research* 10 (1998): 193–200.

Baron, William R. "The Reconstruction of Eighteenth Century Temperature Records Through the Use of Content Analysis." *Climatic Change* 4 (1892): 385–98.

Beerling, D. J. *The Emerald Planet: How Plants Changed Earth's History*. Oxford: Oxford University Press, 2007.

Bektas, M. Yakup, and M. Crosland. "The Copley Medal: The Estab-

lishment of a Reward System in the Royal Society, 1731–1839."
*Notes and Records of the Royal Society of London* 46 (1992): 43–76.

Berkner, L. V., and L. C. Marshall. "On the Origin and Rise of Oxygen Concentration in the Earth's Atmosphere." *Journal of the Atmospheric Sciences* 22 (1965): 225–61.

Berner, Robert A. "The Long-term Carbon Cycle, Fossil Fuels and Atmospheric Composition." *Nature* 426 (2003): 323–26.

Berner, R. A., and D. E. Canfield. "A New Model of Atmospheric Oxygen over Phanerzoic Time." *American Journal of Science* 289 (1989): 333–61.

Boorstin, Daniel J. *The Lost World of Thomas Jefferson.* Chicago: University of Chicago Press, 1993.

Burke, Edmund. *Speech on Conciliation with America.* Edited by A. S. Cook. New York: Longmans, Green, 1896.

Cappon, Lester J., ed. *The Adams-Jefferson Letters.* New York: Simon & Schuster, 1971.

Clark, John Ruskin. *Joseph Priestley: A Comet in the System.* San Diego: Torch Publications, 1960.

Committee on Earth System Science. *Earth System Science: A Program for Global Change.* Washington, DC: NASA, 1988.

Crane, Verner W. "The Club of Honest Whigs: Friends of Science and Liberty." *William and Mary Quarterly* 23 (1966): 210–33.

Cressy, David. "The Vast and Furious Ocean: The Passage to Puritan New England." *New England Quarterly* 57 (1984): 511–32.

Crosland, Maurice. "The Image of Science as a Threat: Burke versus Priestley and the 'Philosophic Revolution.' " *British Journal for the History of Science* 20 (1987): 277–307.

Csikszentmihalyi, Mihaly. *Creativity: Flow and the Psychology of Discovery and Invention.* New York: HarperPerennial, 1997.

Cuvier, Georges. "Biographical Memoir of Joseph Priestley." *New Philosophical Journal,* July–September 1827, vol. iii.

De Vorsey, Louis. "Pioneer Charting of the Gulf Stream: The Contributions of Benjamin Franklin and William Gerard De Brahm." *Imago Mundi* 28 (1976): 105–20.

Dick, Malcolm, ed. *Joseph Priestley and Birmingham.* Studley, England: Brewin Books, 2005.

Dole, Malcom. "The Natural History of Oxygen." *Journal of General Physiology* 49 (1965): 5–27.

Dudley, Robert. "Atmospheric Oxygen, Giant Paleozoic Insects and the Evolution of Aerial Locomotor Performance." *Journal of Experimental Biology* 201 (1998): 1043–50.

Duffy, John. "The Passage to the Colonies." *Mississippi Valley Historical Review* 38 (1951): 21–38.

Ellis, Joseph J. *American Sphinx: The Character of Thomas Jefferson.* New York: Vintage Books, 1996.

———. *Founding Brothers: The Revolutionary Generation.* New York: Vintage Books, 2002.

Eshet, Dan. "Rereading Priestley: Science at the Intersection of Theology and Politics." *History of Science* 39 (2001): 127–59.

Ford, Paul Leicester, ed. *The Works of Thomas Jefferson, Vol. 1–12.* New York: Knickerbocker Press, 1905.

Franklin, Benjamin. *The Life and Writings of Benjamin Franklin, Vol. 2.* Philadelphia: McCarty & Davis, 1834.

———. *The Ingenious Dr. Franklin: Selected Scientific Letters of Benjamin Franklin.* Edited by N. G. Goodman. University Park, PA: University of Pennsylvania Press, 1974.

———. *The Papers of Benjamin Franklin.* Digital edition by the Packard Humanities Institute, sponsored by the American

Philosophical Society and Yale University. http://franklin
papers.org/franklin.

Garrett, Clarke. "Joseph Priestley, the Millennium, and the French
Revolution." *Journal of the History of Ideas* 34 (1973): 51–66.

Gaustad, Edwin Scott. *Sworn on the Altar of God: A Religious Biography of Thomas Jefferson.* Grand Rapids, MI: Wm. B. Eerdmans
Publishing, 1996.

Gay, Peter. *The Enlightenment: The Science of Freedom.* New York:
W. W. Norton, 1977.

Gest, Howard. "Sun-beams, Cucumbers, and Purple Bacteria."
*Photosynthesis Research* 19 (1988): 287–308.

Gibbs, Frederick William. *Joseph Priestley: Adventurer in Science and
Champion of Truth.* New York: Nelson, 1965.

Gohau, Gabriel. "The 26th International Geological Congress,
Paris, 1980." *Episodes* 29 (2006): 123–27.

Golden, Joseph H. "The Life Cycle of Florida Keys' Waterspouts. I."
*Journal of Applied Meteorology* 13 (1974a): 676–92.

———. "Scale-Interaction for the Waterspout Life Cycle. II." *Journal of Applied Meteorology* 13 (1974b): 693–709.

Golden, Joseph H., and H. B. Bluestein. "The NOAA–National
Geographic Society Waterspout Expedition (1993)." *Bulletin of
the American Meteorological Society* 75 (1994): 2281–88.

Govindjee, Krogmann D. "Discoveries in Oxygenic Photosynthesis
(1727–2003): A Perspective." *Photosynthesis Research* 80
(2004): 15–57.

Graham, Jenny. "Revolutionary in Exile: The Emigration of Joseph
Priestley to America 1794–1804." *Transactions of the American
Philosophical Society,* new series, 85, no. 2. (1995): i–xii, 1–213.

Griffith, William P. "Priestley in London." *Notes and Records of the Royal Society of London* 38 (1983): 1–16.

Grubb, Farley. "Morbidity and Mortality on the North Atlantic Passage: Eighteenth-Century German Immigration." *Journal of Interdisciplinary History* 17 (1987): 565–85.

Hausman, Carl R., and A. Rothenberg. *The Creativity Question.* Durham, NC: Duke University Press, 1983.

Henry, Alfred J. "Waterspouts." *Monthly Weather Review* 56 (1928): 207–11.

Holmes, David L. *The Faiths of the Founding Fathers.* Oxford: Oxford University Press, 2006.

Hubert, L. F. "The First Waterspout Discovered on Satellite Photographs." *Monthly Weather Review* 110 (September 1982): 382–84.

Isaacson, Walter. *Benjamin Franklin: An American Life.* New York: Simon & Schuster, 2003.

Jackson, Joe. *A World on Fire: A Heretic, an Aristocrat, and the Race to Discover Oxygen.* New York: Penguin Books, 2007.

Jastrow, Joseph. *Story of Human Error.* Manchester, NH: Ayer Publishing, 1936.

Jefferson, Thomas. *The Writings of Thomas Jefferson,* vols. 1–19. Edited by A. E. Bergh. Washington, DC: Thomas Jefferson Memorial Association, 1905.

——. *The Jefferson Bible: The Life and Morals of Jesus of Nazareth.* Boston: Beacon Press, 2001.

Kelly, Jack. *Gunpowder: Alchemy, Bombards, and Pyrotechnics: The History of the Explosive That Changed the World.* New York: Basic Books, 2004.

Keyes, Charles R. "Review of the Progress of American Invertebrate

Paleontology for the Year 1889." *American Naturalist* 24 (1890): 131–38.

Kramnick, Isaac. "Eighteenth-Century Science and Radical Social Theory: The Case of Joseph Priestley's Scientific Liberalism." *Journal of British Studies* 25 (1986): 1–30.

Kuhn, Thomas S. *The Structure of Scientific Revolutions.* Chicago: University of Chicago Press, 1970.

Lovelock, J. E., and L. Margulis. "Atmospheric Homeostasis by and for the Biosphere: The Gaia Hypothesis." *Tellus* 26 (1974): 2–9.

Lundy, Derek. *The Way of a Ship: A Square-Rigger Voyage in the Last Days of Sail.* New York: HarperCollins, 2004.

Maddison, R.E.W., and F. R. Maddison. "Joseph Priestley and the Birmingham Riots." *Notes and Records of the Royal Society of London* 12 (1956): 98–113.

Manley, Gordon. "The Weather and Diseases: Some Eighteenth-Century Contributions to Observational Meteorology." *Notes and Records of the Royal Society of London* 9 (1952): 300–307.

———. "Temperature Trends in England, 1698–1957." *Theoretical and Applied Climatology* 9 (1958): 413–33.

———. "Central England Temperatures: Monthly Means 1659 to 1973." *Quarterly Journal of the Royal Meteorological Society* (1974): 389–405.

Margulis, Lynn, and Dorion Sagan. *Microcosmos: Four Billion Years of Microbial Evolution.* Los Angeles: University of California Press, 1986.

McCann, H. Gilman. *Chemistry Transformed: The Paradigmatic Shift from Phlogiston to Oxygen.* Norwood, NJ: Ablex, 1978.

Mittelberger, Gottlieb. *Journey to Pennsylvania in the Year 1754.* Translated by C. T. Eben. Philadelphia: John Jos. McVey, 1898.

Morton, Alan Q. "Review: Previous Public Perceptions of Science." *Notes and Records of the Royal Society of London* 48 (1994): 157–59.

Moser, Gerald M. *Seven Essays on Joseph Priestley.* State College, PA: Privately printed, 1994.

Munby, A.N.L. "The Distribution of the First Edition of Newton's 'Principia.'" *Notes and Records of the Royal Society of London* 10 (1952): 28–39.

Osborn, Alex Faickney. *Applied Imagination: Principles and Procedures of Creative Problem Solving.* New York: Scribner, 1963.

Parascandola, John, and Aaron J. Ihde. "History of the Pneumatic Trough." *Isis,* 60, no. 3 (Autumn 1969): 351–61.

Platt, Washington, and R. A. Baker. "The Relation of the Scientific 'Hunch' to Research." *Journal of Chemical Education* 8 (October 1931): 1969–2002.

Priestley, Joseph. "Observations on Different Kinds of Air." *Philosophical Transactions* 62 (1772): 147–264.

———. *The History and Present State of Electricity, with Original Experiments.* London: C. Bathurst and T. Lowndes, 1775.

———. *Experiments and Observations on Different Kinds of Air and Other Branches of Natural Philosophy, Connected with the Subject. . . : Being the Former Six Volumes Abridged and Methodized, with Many Additions.* Birmingham, England: Thomas Pearson, 1790.

———. *An Appeal to the Public on the Subject of the Riots in Birmingham to Which Are Added, Strictures on a Pamphlet, Intitled 'Thoughts on the Late Riot at Birmingham.'* Birmingham, England: J. Thompson, 1791.

———. *Letters to the Inhabitants of Northumberland and Its Neighbourhood, Parts 1 & 2.* Northumberland, England: Andrew Kennedy, 1799.

————. *Lectures on History, and General Policy*. London: Thomas Tegg, 1826a.

————. *Theological and Miscellaneous Works of Joseph Priestley (26 volumes)*. Edited by J. T. Rutt. London: Thomas Tegg, 1826b.

————. *A History of the Corruptions of Christianity*. London: Woodfall and Kinder, 1871.

————. *Memoirs of Dr. Joseph Priestley, Written by Himself.* London: H. R. Allenson, 1904.

————. *Jesus and Socrates Compared*. Whitefish, MT: Kessinger Publishing, 1994.

Priestley, Joseph, and Richard Price. *Sermons*. London: R. Hunter, 1830.

Riley, James C. "Mortality on Long-Distance Voyages in the Eighteenth Century." *Journal of Economic History* 41 (1981): 651–56.

Robinson, E. "New Light on the Priestley Riots." *Historical Journal* 3 (1960): 73–75.

Robinson, Jennifer M. "Lignin, Land Plants, and Fungi: Biological Evolution Affecting Phanerozoic Oxygen Balance." *Geology* 15 (1990): 607–10.

Rutt, John Towill, ed. *Life and Correspondence of Joseph Priestley, Vol. 1 & 2*. London: R. Hunter, 1831.

————, ed. *Memoirs and Correspondence of Joseph Priestley*. Bristol, England: Thoemmes Press, 2003.

Schaffer, Simon. "Priestley's Questions: An Historiographic Survey." *History of Science* 22 (1984): 151–83.

————. "Scientific Discoveries and the End of Natural Philosophy." *Social Studies of Science* 16 (1986): 387–420.

Schofield, Robert E., ed. *A Scientific Autobiography of Joseph Priestley*

*(1733–1804): Selected Scientific Correspondence.* Cambridge: MIT Press, 1966.

———. *The Enlightenment of Joseph Priestley: A Study of His Life and Work from 1773 to 1804.* University Park, PA: Penn State Press, 1997.

———. *The Enlightened Joseph Priestley.* University Park: Pennsylvania State University Press, 2004.

Shapin, Steven. *The Scientific Revolution.* Chicago: University of Chicago Press, 1996.

Shapin, S., S. Schaffer, and T. Hobbes. *Leviathan and the Air-Pump: Hobbes, Boyle, and the Experimental Life.* Princeton, NJ: Princeton University Press, 1985.

Simpson, Joanne, et al. "Observations and Mechanisms of GATE Waterspouts." *Journal of the Atmospheric Sciences* 43 (1986): 753–82.

Standage, Tom. *A History of the World in Six Glasses.* New York: Walker, 2005.

Stanley, Steven M. *Earth System History.* New York: W. H. Freeman, 2004.

Sternberg, Robert J., ed. *Handbook of Creativity.* Cambridge: Cambridge University Press, 1999.

Thorpe, T. E. *Joseph Priestley.* London: J. M. Dent, 1906.

Uglow, Jenny. *The Lunar Men: Five Friends Whose Curiosity Changes the World.* New York: Farrar, Straus and Giroux, 2002.

Van Doren, Carl. *Benjamin Franklin.* New York: Viking, 1938.

Walker, W. Cameron. "The Beginnings of the Scientific Career of Joseph Priestley." *Isis* 21 (1934): 81–97.

Walters, Kerry S. *Benjamin Franklin and His Gods: Beyond Provi-*

*dence and Polytheism*. Champagne: University of Illinois Press, 1999.

Wedgwood, Julia. *The Personal Life of Josiah Wedgwood the Potter*. Revised and edited, with an introduction by C. H. Herford. London: Macmillan, 1915.

Weld, Charles Richard. *History of The Royal Society with Memoirs of the Presidents*. London: John W. Parker, 1848.

Wills, Gary. *Mr. Jefferson's University*. Des Moines, IA: National Geographic Society, 2006.

Wright, John W. "The Rifle in the American Revolution." *American Historical Review* 29 (1924): 293–99.

# INDEX

*Page numbers in italics indicate illustrations.*

# Steven Johnson writes about big ideas.

The world is changing—faster than ever. There are more big ideas and more good ideas out there. And Steven Johnson tells us both where they came from and where they can take us.

His books are insightful, wide-ranging, about the future and about our history. They are essential for business, innovation, technology, history, and science readers.

Bill Clinton gave a talk recently where he discussed some of Steven Johnson's books:

"There's an interesting book—if you want to be optimistic about the future—by Steven Johnson, who's a great science writer. It's called *Future Perfect*. [Two of his earlier] books, one of them is called *The Ghost Map*, which is about how the cholera epidemic was solved in London; and one's called *The Invention of Air*, which is about the discovery of oxygen."

Steven Johnson's curious, dynamic, creative mind reveals a fascinating world of ideas and innovation.

**Steven Johnson has big ideas.**

© Nina Subin

# Everything Bad Is Good for You
## How Today's Popular Culture Is Actually Making Us Smarter

Steven Johnson's hallmark classic on pop culture and technology. In this provocative, unfailingly intelligent, thoroughly researched, and convincing book, Johnson draws from fields as diverse as neuroscience, economics, and media theory to argue that the pop culture we soak in every day—from *The Lord of the Rings* to Grand Theft Auto to *The Simpsons*—is actually sophisticated and, far from rotting our brains, is actually posing new cognitive challenges that are making our minds immeasurably sharper.

"Iconoclastic and captivating." **—The Boston Globe**

"Persuasive...The old dogs won't be able to rest as easily once they've read *Everything Bad Is Good for You*, Steven Johnson's elegant polemic." **—Walter Kirn, The New York Times Book Review**

"Wonderfully entertaining. Steven Johnson proposes that what is making us smarter is precisely what we thought was making us dumber: popular culture." **—Malcolm Gladwell, The New Yorker**

# The Ghost Map: The Story of London's Most Terrifying Epidemic—and How It Changed Science, Cities, and the Modern World

A *New York Times* Notable Book

A riveting page-turner about a real-life historical hero, Dr. John Snow. In the summer of 1854, London has just emerged as one of the first modern cities in the world. But lacking the infrastructure—garbage removal, clean water, sewers—necessary to support its rapidly expanding population, the city has become the perfect breeding ground for a terrifying disease no one knows how to cure. As the cholera outbreak takes hold, a physician and a local curate are spurred to action—and ultimately solve the most pressing medical riddle of their time.

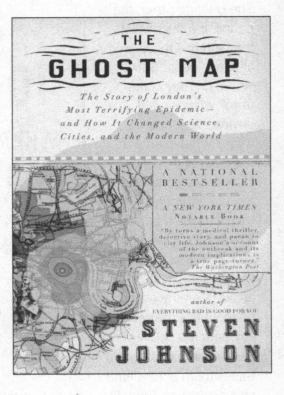

Johnson illuminates the intertwined histories and interconnectedness of the spread of disease, contagion theory, the rise of cities, and the nature of scientific inquiry.

"Thrilling." —*GQ*  "Vivid." —*The New Yorker*

"Marvelous." —*The Wall Street Journal*  "Fascinating." —*The New York Times Book Review*

# Where Good Ideas Come From
## The Natural History of Innovation
A *New York Times* bestseller • An *Economist* Best Book of the Year

The printing press, the pencil, the flush toilet, the battery—these are all great ideas. But where do they come from? What kind of environment breeds them? What sparks the flash of brilliance? How do we generate the breakthrough technologies that push forward our lives, our society, our culture? Steven Johnson's answers are revelatory as he identifies the seven key patterns behind genuine innovation, and traces them across time and disciplines. From Darwin and Freud to the halls of Google and Apple, Johnson investigates the innovation hubs throughout modern time and pulls out the approaches and commonalities that seem to appear at moments of originality.

STEVEN
JOHNSON
"A FIRST-RATE STORYTELLER."—*THE NEW YORK TIMES*
WHERE GOOD IDEAS
COME FROM
A *NEW YORK TIMES* BESTSELLER
THE NATURAL
HISTORY OF
INNOVATION
"ENTERTAINING AND SMART."
—*LOS ANGELES TIMES*
FROM THE BESTSELLING
AUTHOR OF *EVERYTHING
BAD IS GOOD FOR YOU*
AND *THE INVENTION
OF AIR*

"[A] rich, integrated, and often sparkling book. Mr. Johnson, who knows a thing or two about the history of science, is a first-rate storyteller."
—*The New York Times*

"A vision of innovation and ideas that is resolutely social, dynamic, and material... Fluidly written, entertaining, and smart without being arcane."
—*Los Angeles Times*

# The Innovator's Cookbook
**Essentials for Inventing What Is Next**

Innovation is one of
today's buzzwords for
a reason. The need to
push forward, find new
paths and new ideas in
an ever-evolving world, is
a vital part of business,
of education, of politics,
of our daily lives. This
is an essential book
for anyone interested
in innovation: the key
texts on the topic from a
wide range of fields, as
well as interviews with
successful, real-world
innovators, prefaced
with an original essay
from Johnson that draws
upon his own experience
as an entrepreneur
and author.

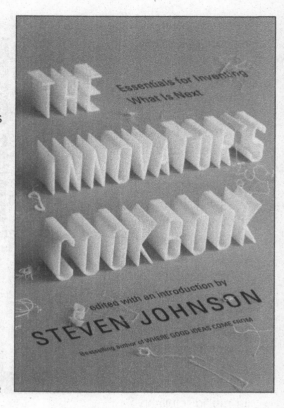

Includes writing from:

| | | |
|---|---|---|
| Stewart Brand | Teresa Amabile | Beth Noveck |
| Clayton Christensen | Peter Drucker | Jon Schnur |
| Richard Florida | Amar Bhidé | Katie Salen |
| | Ray Ozzie | Brian Eno |

# Future Perfect
## The Case for Progress in a Networked Age

Steven Johnson makes the case that a new model of political change is on the rise, transforming everything from local governments to classrooms, from protest movements to health care. Johnson paints a compelling portrait of this new political worldview— influenced by the success and interconnectedness of the Internet, but not necessarily dependent on high-tech solutions—that breaks with traditional categories of liberal or conservative thinking.

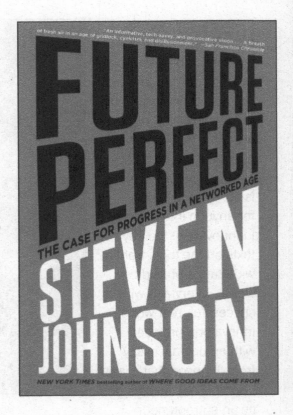

With his acclaimed gift for multidisciplinary storytelling and big ideas, Johnson explores this innovative vision of progress through a series of fascinating narratives. At a time when the conventional wisdom holds that the political system is hopelessly gridlocked with old ideas, *Future Perfect* makes the timely and uplifting case that progress is still possible, and that new solutions are emergent.

"An informative, tech-savvy and provocative vision...a breath of fresh air in an age of gridlock, cynicism and disillusionment." —*San Francisco Chronicle*

"In clear and engaging prose, Johnson writes about this emerging movement...*Future Perfect* is a buoyant and hopeful book." —*Boston Globe*